Innovations in the Memory System

Synthesis Lectures on Computer Architecture

Editors
Natalie Enright Jerger, *University of Toronto*
Margaret Martonosi, *Princeton University*

Founding Editor Emeritus
Mark D. Hill, *University of Wisconsin, Madison*

Synthesis Lectures on Computer Architecture publishes 50- to 100-page publications on topics pertaining to the science and art of designing, analyzing, selecting and interconnecting hardware components to create computers that meet functional, performance and cost goals. The scope will largely follow the purview of premier computer architecture conferences, such as ISCA, HPCA, MICRO, and ASPLOS.

Innovations in the Memory System
Rajeev Balasubramonian
2019

Cache Replacement Policies
Akanksha Jain and Calvin Lin
2019

The Datacenter as a Computer: Designing Warehouw-Scale Machines, Third Edition
Luiz André Barroso, Urs Hölzle, and Parthasarathy Ranganathan
2018

Principles of Secure Processor Architecture Design
Jakub Szefer
2018

General-Purpose Graphics Processor Architectures
Tor M. Aamodt, Wilson Wai Lun Fung, and Timothy G. Rogers
2018

Compiling Algorithms for Heterogenous Systems
Steven Bell, Jing Pu, James Hegarty, and Mark Horowitz
2018

Innovations in the Memory System

Rajeev Balasubramonian

ISBN: 978-3-031-00635-7   paperback
ISBN: 978-3-031-01763-6   ebook
ISBN: 978-3-031-00060-7   hardcover

DOI 10.1007/978-3-031-01763-6

A Publication in the Springer Publishers series
SYNTHESIS LECTURES ON COMPUTER ARCHITECTURE

Lecture #48
Series Editors: Natalie Enright Jerger, *University of Toronto*
                  Margaret Martonosi, *Princeton University*
Founding Editor Emeritus: Mark D. Hill, *University of Wisconsin, Madison*
Series ISSN
Print 1935-3235   Electronic 1935-3243

Innovations in the Memory System

Rajeev Balasubramonian
University of Utah

SYNTHESIS LECTURES ON COMPUTER ARCHITECTURE #48

ABSTRACT

The memory system has the potential to be a hub for future innovation. While conventional memory systems focused primarily on high density, other memory system metrics like energy, security, and reliability are grabbing modern research headlines. With processor performance stagnating, it is also time to consider new programming models that move some application computations into the memory system. This, in turn, will lead to feature-rich memory systems with new interfaces. The past decade has seen a number of memory system innovations that point to this future where the memory system will be much more than dense rows of unintelligent bits. This book takes a tour through recent and prominent research works, touching upon new DRAM chip designs and technologies, near data processing approaches, new memory channel architectures, techniques to tolerate the overheads of refresh and fault tolerance, security attacks and mitigations, and memory scheduling.

KEYWORDS

memory system architecture, DRAM, memory controllers, memory channels, low-power memory, memory security, error correction codes, memory reliability, DRAM refresh, near data processing, 3D stacking

Contents

List of Figures

List of Tables

Preface

The memory sub-system is the most overlooked and underrated component in modern systems. Memory chips are just rows of dumb bits: they are commoditized and standardized; they perform two simple operations—reads and writes.

A lot of the above conventional wisdom is changing. With processor performance improving only slightly with each new generation, the computing industry needs creative value-additions in new systems. Computation is widely viewed as "trivial" or "commoditized," while data movement is viewed as the engineer's nemesis. Most solutions to these problems seem to rely on two broad compelling approaches: accelerators and new memory technologies.

For decades, architecture conferences saw little activity in memory systems. In the past ten years, this area has been a dominant topic, only recently ceding that title to machine learning accelerators. Much of this recent excitement stemmed from new non-volatile memory technologies and 3D stacking. Both made it possible to improve energy efficiency and add more functionality to the memory system. Once the floodgates opened, researchers have delved into several aspects of memory systems, including security, reliability, refresh, microarchitecture, etc.

However, even today, most computer architecture textbooks spend hundreds of pages on the processor and just a handful of pages on the memory system. Most graduate students starting their Ph.D.s in architecture require a solid foundation in memory design. We hope that this Synthesis Lecture serves as a valuable resource to such students, as well as to other architecture researchers/developers. Instructors teaching an advanced computer architecture class may want to base some of their lectures on parts of this book. Chapter 2 would also be appropriate reading for a standard course on computer architecture. To make the material easily accessible to early-stage graduate students, each chapter starts by covering the basics and then diving into key ideas/contributions in recent top-tier papers. The goal here is to provide a condensed big-picture overview of the landscape; once informed, students can chase down specific topics/papers that deserve more attention. While I have read hundreds of papers so you don't have to :-), I highly recommend that you absolutely read the original papers that you find relevant. I also want to recommend the 2008 memory system book written by Jacob, Ng, and Wang [4]; it's a terrific reference that I have frequently consulted.

Given the many popular conferences/workshops in existence today, I have obviously not covered every published memory system innovation. To be fair and realistic, I have primarily covered papers published at the top four architecture venues—ISCA, MICRO, ASPLOS, HPCA. I realize that this approach has shortcomings, and I apologize in advance if I did not mention your favorite work.

The book represents the hard work of a dedicated team at Morgan & Claypool, the careful oversight of series editors Margaret Martonosi and Natalie Enright Jerger, the constructive feedback of reviewers, and the education I have received from my students and collaborators. We hope you enjoy the book!

Rajeev Balasubramonian
July 2019

Acknowledgments

I am very grateful to all my collaborators, especially Al Davis and all my Ph.D. students who have taught me so much about each aspect of the memory system. I can claim to be a memory expert only because my students' intellectual curiosity, creativity, and attention to detail pulled my research program in new and exciting directions. As always, a ton of credit goes to my parents and my highly supportive family—Deepthi, Shurik, and Anushka.

Rajeev Balasubramonian
July 2019

CHAPTER 1

Introduction

Most of us learned the basics of computer architecture from the famous Hennessy and Patterson textbooks [277]. These books clearly emphasize processor and cache design principles. But as the 21st century comes of age, the research playground is shifting away from the processor chip. For the most part, we understand the major principles in defining an in-order core, an out-of-order core, cache replacement policies, etc. The processor cores are therefore being commoditized and the hub of innovation is shifting toward two other major components in modern systems. One is a supporting accelerator like a GPU (Graphics Processing Unit) or TPU (Tensor Processing Unit). The other is the memory system.

In the last decade, there were very few research groups engaged in memory systems research. Every memory system paper included a tutorial that educated readers about ranks, banks, row buffers, etc. Information about commercial memory chips was not readily available. When I entered this research area a decade ago, I found myself piecing together the fundamentals from various sources, and frequently referencing parts of the voluminous memory bible from Jacob et al. [4]. Hopefully, this synthesis lecture serves as a succinct first step for early-stage graduate students, especially if their planned research involves innovation within the memory system.

In the past decade, the landscape has shifted significantly. Many more groups work on memory systems research. Papers aren't required to include a section on Memory 101. Most memory papers have no trouble finding qualified reviewers. In the past, reviewers frequently killed memory papers by invoking, "But the memory industry is very cost- and change-averse. They will never consider ideas like this." Today, that argument carries less weight, although, the mythical Reviewer C is known to occasionally re-appear. While the memory industry continues to be very cost- and change-averse, the door has opened a crack. Academics have rushed in to squint at memory chips, kick the tires, ask "what-if" questions, and generally force complex ideas upon the memory industry. And as with most academic research, some of these ideas will inevitably stick, and bring about small and large changes.

One of the drivers behind the "memory gold rush" was industry's interest in non-volatile memory (NVM) cells. Most researchers have also branched into other aspects of memory design. In fact, we'll leave discussions of NVM to other works [9] and primarily focus on these other memory design aspects in this synthesis lecture: scheduling, data placement, compression, bank microarchitecture, interconnects, reliability, Dynamic Random Access Memory (DRAM) refresh, near data processing, and security. All of these areas have seen significant research activity in the past decade. They have all targeted important bottlenecks, and in some cases, have helped remove these bottlenecks. For example, I consider scheduling and refresh as largely solved problems. For other areas, like near data processing or security, we have only scratched the surface.

In recent years, we have seen the emergence of a number of radically different memory products, e.g., high bandwidth memory (HBM), hybrid memory cube (HMC), and automata processor. While industry had been held to strict DDRx standards by an industry consortium, JEDEC, new interface standards like Gen-Z are also emerging. The time is therefore ripe to re-think memory architectures without the shackles of legacy standards, and define the optimal way to organize a variety of memory products, a variety of memory cells/mats, a variety of interconnects, a variety of data coding/movement policies, etc.

In addition, 2018 was a watershed year for the processor industry. Spectre, Meltdown, and a slew of subsequent attacks have demonstrated that security and privacy cannot be ignored by processor vendors. The memory system is typically outside the trusted computing base, and especially vulnerable to attacks. Therefore, memory system defenses, that had historically been viewed as impractical academic pursuits, are now under serious consideration by industry.

Finally, as we near scaling limits, the memory industry will have to find new ways to add value to their products. The memory package may therefore engage in various auxiliary or maintenance operations (reliability, security, compression, etc.), or include full-fledged processing cores or include accelerators or support various in-memory operators. We have therefore witnessed the second coming of "Processing in Memory," or the more modestly labeled "Near Data Processing." Recent results have shown that such efforts cannot only improve performance and energy for a range of workloads, they can do so with minimal impact on cost.

All of the above-mentioned factors have injected life into the field of memory systems research. The next ten chapters will try to convey the basic elements of each research area, summarize the research landscape, and point the reader to fertile future areas of research.

CHAPTER 2

Memory System Basics for Every Architect

For decades, it was possible to be a computer architect while knowing very little about the memory system. I too stumbled into 2008 not knowing the difference between a *rank* and a *bank*. Even today, many memory system papers include a primer on the memory system because some of their audience has never been formally educated on this topic. However, as explained in Chapter 1, the off-chip memory hierarchy is already the hub for innovation in modern and future systems. *Every* computer architect must therefore understand the basics of memory design, just as he/she understands what's beneath the hood of an out-of-order processor. This chapter therefore presents essential background for every architect. Later, in Section 6.1, we'll dive deeper into DRAM microarchitecture.

2.1 DRAM VS. SRAM

On a processor chip, data is typically stored in Static Random Access Memory (SRAM) caches. However, a SRAM cell is large enough that a single processor chip can only accommodate a few megabytes of data per processing core. Applications typically have memory requirements that run into many giga bytes. To fulfil this memory requirement in a server, a memory system is constructed on separate printed circuit board (PCB) modules using several high-density and low-cost DRAM chips. In mobile devices, the memory system is made up of a few DRAM chips that are typically packaged with the processor in a multi-chip module. Why the focus on high density and low cost for a DRAM chip? Note that if data is not found in memory, a hard disk or Flash drive is typically accessed. Those accesses are far more expensive—many microseconds for Flash or a few milliseconds for hard disk. Memory capacity therefore has a significant impact on overall system performance. Second, the purchase price of a high-capacity memory system is comparable to that of a high-end processor. Third, since we're going off the processor chip[1] to access this memory, high latency is inevitable. So the marginal penalty from higher-latency memory chips should be tolerable, especially if it can significantly improve other system metrics. For these reasons, historically, memory chips have prioritized density over latency. Memory chips are therefore implemented with DRAM cells, a technology that is denser than SRAM. Given the quest for dense memory, industry is also pursuing a slew of new cell technologies,

[1]The literature often shortens "off the processor chip" to "off-chip."

e.g., phase change memory, memristors, and spin torque transfer RAM. Some aspects of these non-volatile memory (NVM) technologies are discussed in a separate synthesis lecture [9].

An SRAM cell uses two back-to-back inverters in a loop to store a bit (see Figure 2.1). Because of this feedback loop, data is retained in the cell in spite of continuous leakage current within the cell. Because of this data retention, the cell is called *Static* Random Access Memory. The use of two inverters per cell also leads to low density per bit.

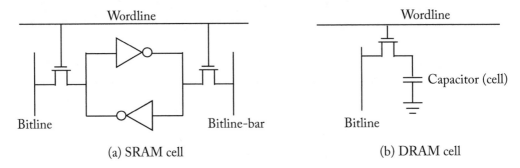

(a) SRAM cell   (b) DRAM cell

Figure 2.1: A comparison of SRAM and DRAM cells. (a) The SRAM cell has two back-to-back inverters and they are connected to two bitlines with two access transistors. Each inverter requires two transistors, so an SRAM cell requires six total transistors, also known as a 6T SRAM cell. (b) The 1T1C DRAM cell only requires one access transistor, one capacitor, and one bitline.

A DRAM cell, on the other hand, is optimized for high density. As shown in Figure 2.1, a DRAM cell is basically comprised of a single capacitor, which is the key to its high density. When charge is stored in the capacitor, it represents a "1", else a "0". Over time, the charge in the capacitor leaks away and the data bit is lost. Hence, to retain the value, it must be read and re-written before the charge is completely lost. Because of this refresh process, the cell is referred to as *Dynamic* Random Access Memory.

2.2   MEMORY CHANNEL

A processor has a memory controller that is responsible for issuing commands to the memory system. The memory chips are largely "dumb;" they are almost entirely controlled by commands emanating from the memory controller. The memory controller is connected to pins on the processor chip, which in turn are connected to a memory channel (wires on a motherboard).

The memory channel is made up of a data bus and an address/command bus. The address/command bus is a unidirectional bus that carries a bits of address and c bits of command from the memory controller to memory chips. The data bus is bidirectional and carries d bits of data from the memory chips to the memory controller for a read request, and vice versa for a write. We won't get into the specifics of the DDR3/DDR4 standard here, but if we reasonably abstract away some of the details, we can move ahead with the assumption that $a = 17$, $c = 5$, and $d = 64$.

The memory channel typically operates at a frequency slower than the processor. The memory system also follows the Double Data Rate (DDR) standard, which uses both the rising and falling clock edges to carry data signals, i.e., a channel data wire carries two bits per cycle. A state-of-the-art channel can operate at a frequency of 1200 MHz. Note that commercial products will say "2400" to capture the DDR nature of the 1200 MHz clock. Note that the address/command bus operates at single data rate.

In order to support a high channel frequency, the channel must be relatively short and must drive a small load. Hence, only a few memory modules can be connected to the channel. Modern high-end processors typically support a handful of memory channels, with each having an independent memory controller.

2.3   DDR STANDARDS

The memory system is designed to be upgradeable, meaning that in most systems one can pull out existing memory modules and replace them with new memory modules. To enable such plug-and-play, the memory interface is governed by the DDR standard defined by an industry consortium JEDEC. Memory vendors like Micron, Samsung, and SK Hynix produce memory chips that abide by the DDR standard, and similarly, processor vendors like Intel, IBM, and AMD produce memory controllers that also fulfil the DDR contract. The existence of the DDR standard makes life easier for both processor and memory vendors, but also introduces inertia that stymies innovation. The DDR standard is updated every few years. In 2016, the industry transitioned from DDR3 to DDR4. While DDR5 product prototypes have been announced in 2019 [6], there is some uncertainty about whether DDR6 will materialize in the future. Each new DDR generation offers higher bandwidth and lower energy; we'll provide more details on this in Section 7.1.1. Given the higher error rate induced by technology scaling, newer DDR generations are also adding more features for reliability, which we will discuss in Sections 3.1 and 8.1.

2.4   DIMMS, RANKS, BANKS, MATS

The memory modules that plug into the memory channel are referred to as *Dual Inline Memory Modules* (DIMMs). Each DIMM is a PCB with memory chips on the front and back. Since memory components follow the JEDEC standard, they are typically interchangeable within a generation, i.e., if you wanted to upgrade the memory in your system, you could pull out a DIMM and replace it with another same-generation DIMM with higher capacity.

A DIMM, shown in Figure 2.2, is organized into 1, 2, or 4 ranks. A *rank* is a collection of DRAM chips that all work together to keep the 64-bit data bus busy in a cycle. If a single DRAM chip has a data input/output width of 8 bits, it is referred to as a ×8 ("by 8") chip. We

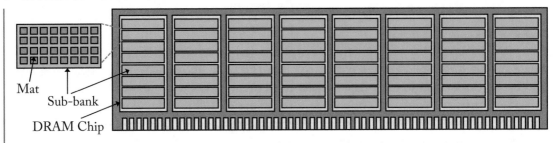

Figure 2.2: Example DIMM: 8 ×8 DRAM chips on one side form 1 rank and the back of the DIMM has another rank. A rank is partitioned into 8 banks. The portion of a bank in one chip can be called a sub-bank. The figure on the left zooms into one sub-bank, which is composed of 32 mats. Each row of mats is called a subarray.

would require 8 such ×8 chips to feed a 64-bit data bus; this collection of 8 ×8 chips is referred to as a rank. You could construct a rank with 64 ×1 chips, 32 ×2 chips, 16 ×4 chips, 8 ×8 chips, 4 ×16 chips, 2 ×32 chips, and 1 ×64 chip (the first two and last two are not commercially popular).

A single data wire on the memory channel is only connected to one DRAM chip pin on a rank. If a memory channel supports four ranks, a data wire must drive four different pins.[2] The number of ranks on a channel is kept small to reduce this load on the bus. On the other hand, an address/command wire on the memory channel is connected to every DRAM chip on every rank on the channel. Because of this higher load, the address/command wires do not use DDR. To reduce this load, special DIMMs called buffered DIMMs are sometimes used. A buffered (registered) DIMM has a buffer chip that receives the address/command signals and then broadcasts them to all the DRAM chips on the DIMM, i.e., an address/command wire on the memory channel only drives one buffer chip per DIMM, not every DRAM chip on the channel.

Fetching data from a rank can take a long time (about 40 ns) because of slow circuits on a DRAM chip. The memory system would be extremely slow if we had to wait for one request to complete before issuing the next. Therefore, multiple requests are serviced in parallel. For a read request, the "memory pipeline" is as follows: the request is issued on the address/command bus; the involved DRAM chips in a rank activate their circuits to pick out the requested data; the requested data is sent back on the data bus. To promote parallelism, once a request has moved on to the second stage of the pipeline, the first stage is available for use by other requests. So while one rank is busy picking out the requested data, we can use the address/command bus to fire requests to other ranks. Therefore, at a time, multiple ranks could be busy picking out the requested data; ultimately, all the data responses will happen sequentially on the single data bus that is shared by the ranks.

[2]Thanks to die-stacked DRAM packages, it is possible for a data wire to drive fewer pins than the number of ranks.

This 3-stage pipeline is not balanced; the first stage (address/command bus) takes about 1 ns, the second stage (data select) takes about 35 ns, and the third stage (data bus transfer) takes about 5 ns. If we want to keep the data bus fully utilized, several requests must be working on the second stage in parallel. The existence of four ranks on a channel helps with this parallelism, but is not enough. Therefore, a single rank is itself partitioned into multiple *banks*. If we assume that a channel has four ranks and each rank is broken into eight banks (as is the case in DDR3), a single channel now supports 32 banks. Each bank can work on its request independently, giving us a high degree of parallelism, and maximizing the chances that some data is ready to be sent on the data bus in every cycle.

If a rank is made up of eight DRAM chips, each bank in that rank also spans across eight DRAM chips. Each bank in this example is therefore partitioned into eight *sub-banks*, where each sub-bank resides entirely in one DRAM chip (sub-bank is a term I made up to reduce confusion). A sub-bank is itself partitioned into many data *subarrays* and *mats* (see Figure 2.2). This is done to reduce the latency for data look-up and to reduce the interconnect overhead. We'll talk more about subarrays and mats later in Section 6.1.

A single 64-byte cache line is fetched from a single bank. The cache line is scattered across all the DRAM chips that form the rank and bank. If we assume a rank that has 8 ×8 chips, it means that each chip (and sub-bank) contributes 64 bits to the entire cache line transfer. Those 64 bits are themselves scattered across multiple mats within that sub-bank. Once the data is selected, the data transfer on the memory channel will require eight 64-bit transfers. For each 64-bit transfer, each chip is contributing eight bits in this example. Keep in mind that the eight 64-bit transfers require four data bus cycles since we're using double data rate.

I realize that this terminology can be a little tedious and off-putting. In essence, the trillions of bits forming a server's memory system are hierarchically organized into DIMMs, ranks, banks, sub-banks, subarrays, and mats. A system administrator configuring a server primarily deals with DIMMs; an architect designing the memory controller or evaluating a system primarily deals with the parallelism afforded by ranks and banks; a microarchitect designing better circuits/wiring/interfaces for a DRAM chip primarily deals with how sub-banks, subarrays, and mats are organized.

2.5   ROW BUFFERS

Tedious terminology notwithstanding, a data request ultimately activates several mats on several DRAM chips. Each mat is simply a matrix of DRAM cells, with say, 512 rows and 512 columns (a 256 Kb mat). The horizontal wire that spans a row is called a wordline. The data request activates a single wordline in each involved mat; all the DRAM cells in that row place their data on vertical wires called bitlines. At the bottom of the mat are sense amplifiers that interpret the signals on the bitlines. It takes about 13 ns to bring a row of data into the sense amplifiers. This is referred to as *Row Activation*, typically initiated with a Row Address Strobe (RAS) command.

Thus, a single cache line request may read out 512-bit wide rows in (say) 64 different mats. The sense-amplifiers therefore store as much as 32 Kb (4 KB) of data. The sense-amplifiers within a bank with valid data are referred to as the *row buffer*. The requested 64-byte cache line is then selected out of this 4 KB row buffer and sent back on the memory channel, with each mat contributing eight bits in this example. This is initiated with a Column Address Strobe (CAS) command. A cache line request therefore leads to "overfetch" of data into the row buffer. The row buffer can serve as a cache inside the DRAM chips, storing 64 different cache lines in our example. If the next cache line request is already present in the row buffer, it can be serviced at a lower latency and energy cost. Only a single row can be active in a bank at a time. A discussion on microarchitecture details has been deferred until Section 6.1.

Before a new row can be read out of a bank, the bitlines have to be precharged. This step, that readies the bank for the next access, also takes about 13 ns. As soon as the bitlines are precharged, the contents in the sense-amplifiers (the row buffer) are lost.

This leads to three types of memory accesses. The most favorable case is a row buffer hit, where the requested cache line is already in the row buffer. Such an access has a latency of about 13 ns, which is the cost of shipping the cache line from the mat to the DRAM chip output pins. The second case is an empty row access, where the bitlines have been precharged in advance and the row buffer stores nothing. Such an access takes 26 ns (13 ns for the row activation that populates the row buffer and 13 ns for the transfer to the output pins). The third case is a row buffer conflict, where a different row currently occupies the row buffer. This access takes about 39 ns: 13 ns to precharge the bitlines, 13 ns to read the new row, and 13 ns for data transfer to output pins. It is the responsibility of the memory controller to issue precharges at the right time so that row buffer hits and empty row accesses can be maximized.

Once a Last Level Cache (LLC) miss has been detected, the resulting memory access can take over 100 ns (300 processor clock cycles at a 3 GHz frequency)—say, 60 ns of queuing delay at the memory controller, 39 ns for a row buffer conflict access, and 4 ns for the transfer on the data bus.

2.6 CAPACITY VS. ENERGY

We've seen that a cache line request activates all the chips in the rank and many mats within these chips. Each mat is made large to reduce the overheads of the peripheral circuits for that mat (decoders, sense amplifiers). This leads to wide rows and significant overfetch. This overfetch incurs a high energy penalty, but is done to maximize density and reduce cost-per-bit, which is a primary metric for DRAM chips.

If we use wide-output DRAM chips to construct a rank, say 4 ×16 2 Gb chips, we can service a cache line by activating few chips, thus reducing overfetch and energy. But such a rank would only have a capacity of 8 Gb. On the other hand, if we used narrow-output DRAM chips, say 16 ×4 2 Gb chips, we would increase activation energy and overfetch, but a single rank would now support 32 Gb capacity. These trade-offs must be considered when populating a memory system with the appropriate set of DIMMs.

2.7   ADDRESS MAPPING

The memory controller can interpret addresses in different ways, each leading to a different placement of data in memory. The assumption is that applications will exhibit data locality and access consecutive cache lines within a short time window. To promote row buffer hits, consecutive cache lines can be placed in the same row. To promote parallelism, consecutive cache lines can be placed in different ranks and banks. For the first mapping policy, the data address can be interpreted as follows:

$$row : rank : bank : channel : column : blkoffset$$

The last few bits represent the data offset within the cache line. The next few bits represent different column bits within a row buffer. Consecutive cache lines will first differ in these column bits, thus representing different elements within a single row buffer. The second mapping policy interprets the data address as follows, where the last few fields represent different channels, banks, and ranks:

$$row : column : rank : bank : channel : blkoffset$$

2.8   SCHEDULING

The memory controller incorporates all the intelligence in the memory system. It is responsible for issuing commands in a manner that optimizes performance, energy, and fairness. The following considerations play a role in determining a good schedule.

First, as discussed above, the scheduler must try to keep a row open so it can service row buffer hits. This is referred to as an open-page policy and works well for applications that have high data locality. However, an open-page policy will eventually lead to an expensive row buffer conflict when it is time to access a new row. If an application has little locality, it is better to use a close-page policy that precharges the bitlines immediately after a cache line and other enqueued row buffer hits have been serviced. This reduces row buffer hits and row buffer conflicts, and leads to more empty row accesses. Modern memory controllers use proprietary algorithms that implement a policy somewhere between an open- and close-page policy.

Second, the direction of the data bus must be switched every time we alternate between a read and a write. This is referred to as bus turnaround time (about 7.5 ns) and must be minimized. Hence, reads and writes are typically serviced in bursts. Reads are always prioritized and serviced first. Writes are not as critical (since a write to memory is generated when a dirty block is evicted from cache) and are placed in a write buffer. When the write buffer is close to filling up (reaches a high water mark), we switch the bus direction and start draining writes until it reaches a low water mark. At that point, the bus direction is switched again and we resume servicing reads.

Third, the scheduler must examine the queue of pending reads or pending writes every cycle to pick out an appropriate read or write that maximizes performance. Often, a scheme First Ready, First Come First Served (FR-FCFS) is used that prioritizes potential row buffer hits over other requests, i.e., exploit the row buffer locality before the bank moves on to another row. Note that the memory controller is shared by several cores on a chip. Since FR-FCFS may end up prioritizing a thread with high row buffer locality, additional mechanisms may be required to enforce fairness.

Fourth, the scheduler must issue refresh operations at the right time. In the worst-case, a DRAM cell is expected to lose its charge within 64 ms. When a row is activated, the cells in that row are automatically refreshed. Since not all rows are activated within every 64 ms window, a background refresh process is required that sequentially steps through every row in the memory sytem and refreshes that row. The 64 ms window is partitioned into 8,192 smaller windows, each lasting 7.8 μs. In each 7.8 μs window, the memory controller issues a refresh command that keeps the memory system busy for a few hundred cycles while it refreshes a few rows in every bank and every rank on the channel. We discuss the refresh process in detail in Section 9.1.

2.9   DRAM TIMING PARAMETERS

We'll first discuss the basic operations involving a DRAM cell. The precharge step brings the voltage of the bitline to $V_{dd}/2$. When a wordline is activated, each DRAM cell in that row connects to its bitline. Depending on the cell content, the bitline voltage either bumps up or down in a process called *charge sharing*. Because of the charge sharing process, the original charge in the cell is lost, i.e., the cell read is destructive. The sense-amplifier detects if the bitline voltage is now $V_{dd}/2 + \delta$ or $V_{dd}/2 - \delta$. It then amplifies the bitline voltage to the cell content, either V_{dd} or 0. Because of this amplification, the cell's state is also restored.

Given the above process, certain timing restrictions have to be obeyed when issuing DRAM commands. In addition, DRAM chips have charge pumps that help raise voltages and supply power; some operations like row Activation deplete the charge pumps; more restrictions are imposed on DRAM commands to allow enough time to replenish charge pumps. DDR chips are therefore defined by a slew of timing parameters; some of the key ones are summarized in Table 2.1, including representative values for DDR3 chips. The time taken to Activate a row, i.e., the time required for the sense-amplifier to sense the cell contents is tRCD, the aforementioned 13 ns delay. tRP and tCAS are the other two 13 ns delays discussed earlier: the time to precharge a bank and the time to move data from row buffer to pins, respectively. The bank can be precharged only after the cells have been restored, which takes a little longer—this is the tRAS delay. tRC is another commonly used DRAM timing parameter—it determines the minimum delay between accesses to different rows in a bank.

You are under no obligation to memorize these timing parameters. Print out Table 2.1 as a handy cheat sheet. Feel free to casually mention tFAW at the next grad student party to establish your very coveted social status as a DRAM guru.

Table 2.1: DRAM timing parameter descriptions [3, 4] *Continues.*

Timing Parameter	Typical Value (cycles at 800 MHz DDR3)	Description
tRCD	11	Row to Column command Delay. This is the time taken to Activate a row. It is the minimum gap between issue of the Activate or RAS command and the subsequent Column-Read or CAS command that identifies a cache block within the row buffer.
tRP	11	Row Precharge. The minimum gap between a precharge command and the subsequent Activate (RAS) command.
tCAS or tCL	11	Column Address Strobe latency. This is the gap between the CAS command and the start of the data transfer on the memory channel.
tRC	39	Row Cycle. This is the minimum gap between accesses to different rows in a bank. For correct operation, tRC =tRAS +tRP.
tRAS	28	Row Address Strobe. This is the time between the start of an Activate operation and data restoration in the cells. The memory controller waits at least tRAS cycles before issuing a precharge command to that bank.
tRRD	5	Row activation to Row activation Delay. This is the minimum gap between two row activations to different banks in a rank. The Activate draws a significant amount of current and the DRAM chip must throttle the rate of Activations.
tFAW	32	Four Activation Window. Similar to tRRD, this is used to throttle the rate of Activations and control peak current. No more than four Activate commands can be issued in any tFAW-cycle window.
tWR	12	Write Recovery time. When performing a write, the memory controller has to provide enough time for the cells to be correctly restored with the just transmitted data. This is the minimum gap between the end of the write data burst and a subsequent precharge command to that bank.
tWTR	6	Write To Read delay. The minimum gap between the end of a write data burst and the start of a subsequent CAS command.
tCWD	5	Column Write Delay. The gap between the issue of a column-write (CAS) command and the start of the write data burst.

Table 2.1: *Continued.* DRAM timing parameter descriptions [3, 4]

tRTP	6	Read to Precharge. The minimum gap between a CAS and a precharge command to that bank.
tCCD	4	Column-to-Column Delay. The minimum gap between two CAS commands to the same bank.
tBURST	4	Data transfer time (64 bytes) on the memory channel.
tRTRS	2	Rank-to-rank switching time. The minimum gap between commands to different ranks. This gap is required so that bus termination can be correctly set up.
tRFC	128	Refresh Cycle time. Other commands can resume only tRFC time after a refresh command has issued.
tREFI	6,240	Refresh interval period. The expected average gap between two refresh commands. The memory controller is allowed some slack in scheduling refreshes.
tPDMIN	4	Minimum power down duration.
tXP	5	Time to exit fast power down.
tXPDLL	20	Time to exit slow power down.

CHAPTER 3

Commercial Memory Products

By now, you've heard this multiple times: DRAM-based memory systems are commodities, enabled by standards like DDR. But standards also kill innovation—they mandate one specific approach for connecting processors and memory. This was perfectly acceptable when the processor was the hub of innovation and the memory simply had to offer low cost-per-bit. But as processor innovations taper out and as the role of memory grows, there is a demand for innovation in memory. We are therefore seeing a proliferation in memory products, many that abide by DDR standards, and a few that create new interfaces.

A complete overhaul of the memory system with non-standard parts would be very expensive. Instead of disrupting (i) the memory controller, (ii) the interconnect architecture, (iii) the memory package, and (iv) the DRAM chip, most innovations will only modify a subset of this list to keep cost under control. This chapter will first discuss a vanilla memory system, and then discuss other commercial products that gradually disrupt more components of the memory system.

3.1 BASIC DDR3/DDR4 CHANNELS AND DIMMS

A majority of processors support one or more memory channels that abide by the DDR3 or DDR4 standard. In a conventional DDR3 memory system, a memory controller on the processor is connected to dual in-line memory modules (DIMMs) via an off-chip electrical memory channel (see Figure 3.1). Modern high-performance processors have four to six memory controllers and memory channels [10–12]. The processor pin counts are not expected to change much as they have neared scaling limits [13]. A modern DDR3 channel has a 64-bit data bus and a 23-bit address/command bus. In order to boost processor pin bandwidth, the memory channel operates at higher frequencies in every new generation. This can significantly increase power consumption; to keep the power in check, the supply voltage and voltage swing have also been lowered. For example, the typical supply voltages for DDR3, DDR4, and initial DDR5 offerings is 1.35 V, 1.2 V, and 1.1 V, respectively [2, 6, 7]. The higher frequency for the channel also forces each new memory generation to support fewer DIMMs per channel. Thus, it is increasingly harder to simultaneously support higher memory bandwidth and higher memory capacity with traditional approaches.

RDIMMs and LRDIMMs Instead of routing wires directly from the memory controller to every DRAM chip on a DIMM, a buffer chip on the DIMM can serve as an interface between

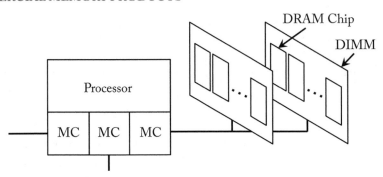

Figure 3.1: An example processor with three memory controllers and a basic DDR3 memory system with two DIMMs per channel.

the memory controller and DRAM chips. Such DIMMs place a much smaller load on the memory channel. Thus, a traditional DDR3 memory channel is able to support many more DIMMs and a higher memory capacity. In an RDIMM (Registered DIMM [14]), shown in Figure 3.2a, the buffer chip (or register chip) only receives address/command signals from the channel and re-distributes each signal to every DRAM chip on the RDIMM. The data wires are directly routed to the DRAM chips without going through the buffer chip. In an LRDIMM (Load Reduced DIMM [15]), shown in Figure 3.2b, viewed as the DDR3 gold standard, both address/command and data signals are first routed to the buffer chip. An LRDIMM therefore places a lower load on the channel's data signals than an RDIMM. The buffer chip on RDIMMs and LRDIMMs is much simpler than the SMB or AMB (discussed shortly) because it does not have to deal with Serialization-Deserialization (SerDes) signaling or converting signals between two protocols/frequencies. Because the channels operate as regular DDR channels, the use of RDIMMs and LRDIMMs does not improve processor pin bandwidth—it only improves the number of total DIMMs that can be supported by a channel at a given frequency.

The DDR4 LRDIMM (Figure 3.2c) goes one step further. To reduce interconnect lengths and improve channel frequencies, the data bits are not routed through a central buffer chip, but through many smaller buffer chips scattered all over the DIMM. The address/command signals continue to use the central buffer chip.

DDR5 Changes   While the DDR5 specification isn't fully defined at the time of writing, the following changes are expected [7, 16]. Frequencies and effective bandwidth will be higher, with bus frequency up to 3.2 GHz (what is referred to as 6400 MT/s with DDR); this is twice as much as the highest bus frequency expected with DDR4. Similarly, the number of bank groups will also be doubled for higher parallelism. While DDR4 introduced fine-granularity refresh that refreshed half the rows in all the banks, DDR5 is expected to make the granularity even finer, similar to LPDDR2's per-bank refresh (some of these refresh approaches are discussed in Section 9.1). Notably, a DDR5 DIMM will be partitioned into two 40-bit channels, each

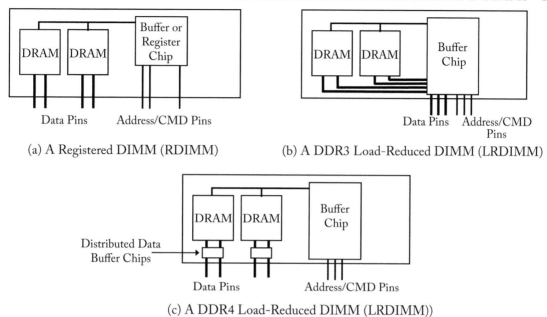

(a) A Registered DIMM (RDIMM)

(b) A DDR3 Load-Reduced DIMM (LRDIMM)

(c) A DDR4 Load-Reduced DIMM (LRDIMM))

Figure 3.2: Organization of RDIMM, DDR3 LRDIMM, and DDR4 LRDIMM. The RDIMM only reduces load on the address/command bus. The DDR3 LRDIMM reduces load on both data and address/command buses. The DDR4 LRDIMM uses additional data buffer chips to reduce the data wiring overheads.

requiring a longer burst length for a full 64-byte transfer. This is motivated by the many research papers that have shown the higher parallelism and energy efficiency from narrower ranks [17–19]. This approach also increases the number of error correction bits per data block, relative to prior DDR versions that allocated 8 bits of error correction for every 64-bit channel transfer. Combined with in-DRAM ECC, this highlights the need for efficient reliability techniques in future technologies.

GDDR5 and Beyond   GPUs typically use GDDR5 memory chips that are optimized for high bandwidth. Because these chips are placed directly on the GPU card, i.e., without using the DIMM stubs that are typical in DDR3/DDR4, the channels are capable of much higher frequencies, offering up to 8 Gbps in GDDR5, up to 12 Gbps in the next-generation GDDR5X, and up to 16 Gbps in the upcoming GDDR6 [5, 20]. As with DDR, new generations of GDDR operate at lower voltages. GDDR standards also allow lower granularity data fetches of 32 bytes [5]. Table 3.1 summarizes key parameters for various DDR and GDDR generations.

Table 3.1: Summary of DRAM parameters across generations [2, 5–7]

Parameter	DDR3	DDR4	DDR5	GDDR5	GDDR5X	GDDR6
Voltage	1.35 V	1.2 V	1.1 V	1.35 V	1.35 V	1.25 V
Frequency (up to)	2.1 Gb/s	4.8 Gb/s	6.4 Gb/s	8 Gb/s	12 Gb/s	16 Gb/s
Access granularity	64 B	64 B	32 B	32 B	32 B	32 B

3.2 DDR DEVIATIONS FOR HIGHER CAPACITY AND BANDWIDTH

We next describe techniques that have been considered to boost memory capacity and bandwidth per processor chip. Nearly all of these innovations have been spearheaded by industry, and very few academic studies have quantified the behavior and relative strengths of each approach.

FB-DIMM In an effort to boost memory capacity per channel, buffered segmented channels, such as FB-DIMM [21], have been proposed. These channels have not been popular because of their higher latency, cost, and power overheads. FB-DIMM [21–23] employs a daisy-chained set of high-speed point-to-point buses (see Figure 3.3a). Each link has uni-directional buses in either direction. Two consecutive buses are connected via a buffer chip on a DIMM called the Advanced Memory Buffer (AMB). The buffer chip is also connected to memory chips on that DIMM. Requests originate at the memory controller and sequentially hop from one buffer to the next. The requests are in a special packetized format and the AMB converts them into regular DDR2 or DDR3 commands when accessing the DRAM chips on the DIMM. The memory controller at the processor is responsible for issuing the appropriate commands and meeting the resulting conservative timing constraints. Similarly, data responses hop sequentially back from the DIMM to the memory controller. Since the AMB can't handle collisions nor buffer packets, it is the responsibility of the memory controller to ensure that commands and responses are sent at appropriate times. FB-DIMM has support for fixed latencies or variable latencies, based on the number of hops [22]. However, to the best of our knowledge, the variable latency option was never used commercially because of the complexity it introduces at the memory controller. The FB-DIMM channel is narrow and employs high-speed SerDes signaling to reduce data transfer time. FB-DIMM has high power overheads because of the faster channel frequency, the need to repeat signals at every hop, perform serialization-deserialization (SerDes), and convert packetized commands into DDR commands. It has been reported that the AMB chip consumes 3-4 W, while each DRAM chip only consumes 0.3 W [24]. We'll discuss SerDes and other links in more detail in Chapter 7.

Buffer-on-Board (BoB) The Buffer-on-Board (BoB) solution with a Scalable Memory Buffer (SMB) chip and a Scalable Memory Interface (SMI) interconnect [25–27] was developed by

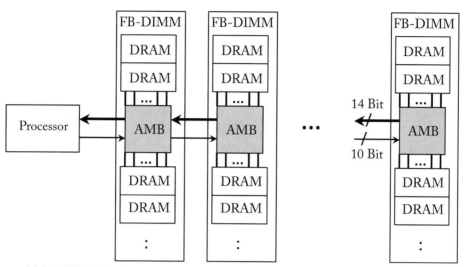

(a) FB-DIMM architecture; up to 8 FB-DIMMs can be daisy-chained through AMBs.

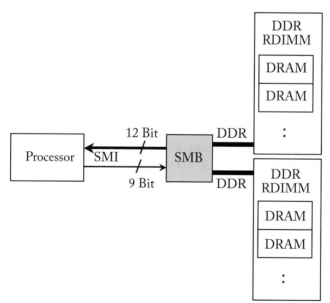

(b) SMI/SMB architecture; at most one SMB and two DDR channels per SMI.

Figure 3.3: Intel's FB-DIMM and Buffer-on-Board (with SMI/SMB) architectures for higher memory capacities.

Intel and represents a more recent incarnation of the FB-DIMM architecture. IBM also has a very similar System Memory Interface (SMI) chip [28] that is used in various Power systems. The BoB solution is used to boost capacity and bandwidth in modern machines, while also reducing trace complexity on the motherboard. As shown in Figure 3.3b, a narrow high-speed SMI SerDes channel connects the processor's memory controller to an SMB chip. The SMB is then connected to two regular DDR3 channels that can each support two DIMMs. Since the SMB multiplexes two DDR3 channels into one narrow SMI, the SMI has to operate at a much higher frequency than the DDR3 channels. Thus, 21 pins on the processor chip, operating at high frequency, are able to support up to four DIMMs. The BoB solution is therefore able to significantly boost processor pin bandwidth and memory capacity. However, like the AMB, the SMB chip is responsible for gathering and serializing data from its DDR3 channels. It therefore introduces non-trivial latency, power, and verification overheads [29]. The SMB datasheet reports nearly 14 W of power dissipation at moderate utilization [25, 29]. Unlike the AMB, the SMB cannot be daisy-chained to yield larger capacities.

3D Memory Stacks One of the most compelling approaches to boost capacity in the near-term is the use of 3D stacking of memory chips. Already, several prototypes and products are available from various companies [1, 30–32], with up to eight chips on a stack. The single memory chips on a DIMM can be replaced by 3D memory chip stacks to form high-capacity DIMMs. The use of stacking does not place additional load on the memory bus, although, it will increase load and energy on some interconnects within the stack. On a 3D stack, each memory chip typically operates independently. Thus, logically, the 3D stack is akin to a large-capacity, many-bank DRAM chip.

3D Memory+Logic Stacks Micron has pioneered the design of a 3D-stacked memory+logic device, the hybrid memory cube (HMC), that has many unique features [1, 33]. The HMC has been designed to deliver high performance and low energy, especially for the high performance computing (HPC) segment. As shown in Figure 3.4, the HMC has 4-8 DRAM chips and a logic chip in a single 3D-stacked package. Each DRAM chip is partitioned into 32 banks to provide low latency and low energy. An entire cache line is fetched from a single bank on a single DRAM chip in the stack, contributing greatly to the HMC's energy efficiency. This cache line is sent to the logic chip via Through-Silicon Vias (TSVs). From the logic chip, the data is sent to the processor via one of eight SerDes links. Each link is made up of 16 upstream and 16 downstream signals operating at very high frequencies. HMC breakdowns show that the energy per bit for an HMC is only 10.48 pJ, of which 3.7 pJ is in the highly efficient DRAM access, and 6.78 pJ is in the logic layer. The HMC's eight links can be used to connect to the processor or to other HMCs, thus opening up the possibility for a network of memory chips on the motherboard. At Supercomputing 2013, Fujitsu exhibited a board that has three processor sockets and eight HMC devices per socket [34]. In this board, each HMC is directly connected

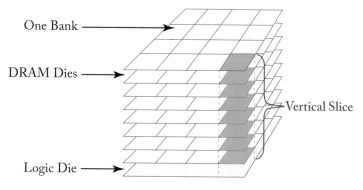

Figure 3.4: The Hybrid Memory Cube (HMC) organization. Based on T. Pawlowski, Hot Chips 2011 [1].

to a processor socket. The logic layer can accommodate various functionalities, although not much is known about the expected roadmap for Micron or other memory vendors.

High-Bandwidth Memory (HBM) and Interposers   In addition to the memory devices mentioned above, there is also a push to stack memory dies on top of a processor to implement (say) a DRAM cache. The HBM JEDEC specification [35], advocated by AMD and SK Hynix, is expected to define the connections between such a memory stack and the processor. There is also a push to use silicon interposers to connect multiple devices, a cheaper so-called 2.5D alternative to 3D stacking. A passive silicon interposer is a relatively cheap piece of silicon that has no transistors, but multiple layers of metal. A processor chip can connect to metal wires on the interposer with a large number of microbumps, thus increasing the bandwidth emerging from a processor chip. Traditionally, a processor connects to the motherboard with C4 bumps that have larger diameters and are more limited in number. The metal wires on the interposer can route signals from the processor to other devices on the same interposer. One example is to use the interposer to connect the processor to a few 3D-stacked DRAM chips that serve as main memory or as a large cache. HBM2 products are available today and HBM3 products are expected to be available in 2020.

CHAPTER 4

Memory Scheduling

A central feature in commodity DDR-based systems is that the memory controller is the master and the memory chips on a DIMM are slaves. The memory controller does not simply provide an address and wait for a response—it actually controls all aspects of the operation, scheduling low-level commands, while being aware of the time for every sub-operation and the state of each DRAM bank.

Why was this choice made? It certainly reduces the burden on memory vendors and they can focus on cost-optimized dense chips. It also retains more performance control within the processor. One can make a valid technical argument that the processor should simply provide an address, leave the low-level control to the memory device, and simply wait for a response to show up. In fact, such an interface is used for the Micron HMC. Business considerations ultimately drive the interface that is selected.

In this chapter, we'll examine how the memory controller schedules low-level commands to the memory device. This controller is typically on the processor that drives the DDR memory channel, but similar principles may also be employed to design a controller that sits within the memory device, as is done for the HMC.

4.1   MEMORY SCHEDULER BASICS

We'll first build on some of the basics of memory scheduling already discussed in Section 2.8. On an LLC miss, a read request for a cache line is typically sent to the memory controller. If the incoming cache line displaces a dirty line in the LLC, that dirty line is sent to the memory controller as a write request. The memory controller has a read queue and a write queue to accommodate these incoming read/write requests.

A canonical memory controller is free to reorder the requests in these read and write queues. Based on a number of heuristics (which is the subject of most of this chapter), one of these requests is selected as highest priority and moved into per-bank command queues. When a read/write request is selected, the memory controller is aware of the preceding state of the bank; it accordingly decomposes the request into multiple commands which are then placed in that bank's command queue. For example, if the request is to an open row, then a single column-read (CAS) command is enough to service a read request. Instead, if the request causes a row conflict, then three commands are inserted into the command queue: precharge, activate (RAS), column-read (CAS). The memory controller is also responsible for placing various other commands in the command queue that may not correspond to read/write requests, for example, precharge (to implement a close-page policy), refresh, power up/down.

Once commands are placed in the command queue, they cannot be re-ordered; note that a command was determined based on the bank's preceding state, so those dependences cannot be violated. For the head of each FIFO command queue, timing constraints are enforced by the memory controller. For example, once an activate is issued from a command queue, the memory controller remembers that the next head of that queue (a column-read command) cannot be issued until tRCD has elapsed. Similarly, if the memory channel is currently servicing a transaction from another rank, the column-read must ensure a gap for rank-to-rank switching (tRTRS). Thus, after enforcing several timing constraints, the heads of each per-bank command queue are flagged as being ready or not. One of these commands is then selected, typically with a fair or round-robin scheme, and issued on the memory channel through a physical interface (referred to as PHY in the literature).

The policies in the memory controller that have the biggest impact on performance are the ones used to populate the command queues. We already discussed some of these considerations in Section 2.8. Is a row buffer hit available in the read or write queues? Are more row buffer hits likely or should a page be closed (should the bank be precharged)? Should the next request be a read or a write? Is now a good time to issue a refresh? We'll next discuss state-of-the-art memory controllers that try to optimally answer the above questions, while also managing thread priorities.

4.2 EARLY MULTI-CORE MEMORY SCHEDULERS

Significant steps were taken in memory scheduling in 2007–2010. This progress was driven by the observation that the gold standard scheduler, FR-FCFS [36], was designed in the single-core era. In a multi-core processor, the handful of memory channels are shared by tens of cores; the memory controller that drives each channel must therefore cater to the demands of many cores/threads.

Fair Queuing Memory System and STFM

The Fair Queuing Memory System [37] and Stall-Time Fair Memory [38] (STFM) schedulers are based on the observation that FR-FCFS prioritizes row buffer hits, thus prioritizing applications that have more locality and more row buffer hits. This leads to unfair treatment of threads, potentially lowering the quality of service (QoS). This is also detrimental for overall system throughput (measured in terms of the weighted speedup of threads). Nesbit et al. [37] scale a thread's timing parameters in proportion to its memory bandwidth allocation so the thread does not take more than its fair share of resources. Mutlu and Moscibroda [38] improve QoS and throughput by first defining the slowdown of each thread by computing the ratio of actual execution time and the estimated execution time without any interference from other threads. If the ratio of maximum and minimum slowdowns experienced by the many threads exceeds a threshold, they prioritize the threads with highest slowdown. This ensures that some threads do not suffer inordinately high slowdowns. Not coincidentally, overall throughput also improves—the marginal utility of resources assigned to a "starved" thread is likely high.

PAR-BS

Mutlu and Moscibroda then followed with a new scheduler, PAR-BS [39]. It handles QoS and fairness by creating batches of requests, and prioritizing older batches, thus preventing a thread from monopolizing the memory controller. Additional rules are created within a batch to maximize throughput: first prioritize row buffer hits, then prioritize threads with a higher "rank," then prioritize requests by age. Threads with low memory intensity receive a higher rank. The use of such a rank also ensures that a thread's requests are potentially serviced in parallel, thus reducing the thread's wait time.

TCM

Kim et al. [40] improved upon PAR-BS with a Thread Cluster Memory (TCM) scheduler. Threads are first organized into separate latency-sensitive and bandwidth-sensitive clusters based on their memory intensity. The latency-sensitive cluster is given higher priority because those threads risk a high slowdown otherwise. Within the bandwidth-sensitive cluster, priority is based on "rank." This rank is determined based on the "niceness" of a thread. A thread with low row buffer hit rate and high bank level parallelism is considered "nice" to others. The rank for a thread is periodically shuffled with insertion shuffling or random shuffling. Insertion shuffling is more effective if there is a big gap in niceness among threads.

Memory Scheduling Championship

The above policies exploit most of the potential benefit with smart memory schedulers. A Memory Scheduling Championship was conducted at ISCA 2012 to evaluate various policies with a common evaluation framework. Most of the submissions added a few new ideas to the principles discussed above, and tuned the policies to eke out a few percentage points of improvement. In fact, most of the submissions achieved performance that was within 2% of each other, and about 7% better than a basic scheduler; there has been little advancement in this space since then. An analysis of the submissions shows that the key considerations in these schedulers are: row buffer locality, read/write alternation, early precharge of a bank when possible, early activation of a bank when possible, hiding refresh, promoting fairness, and managing power-down modes when dealing with a low-energy memory system. An ideal scheduler stitches together several of these techniques to create a "Franken-scheduler;" each technique typically contributes less than a percentage point of improvement. Another take-away from the championship was that smart handling of read/write interference is the single biggest contributor to performance improvement. Surprisingly, none of the championship schedulers employed machine learning; to date, only a couple papers [41, 42] have leveraged reinforcement learning, with some help from genetic algorithms [42]. Given recent advances in both machine learning algorithms and hardware for machine learning, perhaps it is worth revisiting this approach to discover a potential sweet spot in terms of performance and complexity.

BLISS

Most recently, the BLISS scheduler [43] was shown to achieve higher performance and better fairness than many of the prior schedulers discussed here. It also focuses on low complexity. Both features—low complexity and better fairness—are made possible by avoiding having to maintain per-thread ranks and a total order among threads. Instead, applications are split into just two groups—an interference-causing group and an interference-vulnerable group. The scheduler then prioritizes the vulnerable group over the interference-causing group. The classification into groups is done with a simple heuristic that counts the number of consecutive requests serviced for a thread in a given time window; a high count indicates that the thread is interference-causing. The classification information is cleared periodically (every 10K cycles). Within a group, row buffer hits and age are used for priority.

PARDIS

Bojnordi et al. [44] observe that different scheduling heuristics and address mapping schemes may be optimal for different benchmarks. To enable a programmer to dictate the optimal memory scheduler for a workload, the authors design a programmable memory controller. The controller has three stages: a request processor that implements the address mapping (Section 2.7), a transaction processor that implements the scheduling heuristics discussed in this section, and the command logic that implements the timing constraints. While the first two stages are programmable, the last stage is not. The last stage ensures correct timing behavior, while the first two stages impact latency, throughput, and QoS.

4.3   CO-DESIGNED SCHEDULERS

A few other papers have introduced schedulers that are designed to work well with other architecture components and modifications.

Co-Design with LLC

The virtual write queue (VWQ) [45] views the last level cache as a large potential write queue for the memory controller. When the memory controller is draining writes from its modestly sized write queue, it looks at the LLC to find other nearby blocks that could be evicted out of the LLC and efficiently written together to the memory with row buffer hits. It thus performs eager LLC writeback in a manner that improves the memory controller's write drain process.

Co-Design with DRAM Chip

In a related vein, the Staged Read architecture [46] tries to hide read latency by performing partial read operations while a write drain is being performed. In a conventional DDR system, reads must wait for a burst of writes to be performed; this avoids the latency of turning around the direction of the memory bus when alternating between reads and writes. In the Staged Read design, while the write drain is being performed, a number of read requests are also issued. The read data is buffered on the DRAM chip so they can be quickly returned after the write burst. The addition of these buffers to the DRAM chip causes a slight increase in area though.

Co-Design with Prefetcher

Prefetchers also introduce non-trivial interactions within the memory controller. Lee et al. [47] show that some programs benefit by prioritizing demand fetches over prefetches within the memory controller, while other programs benefit from the opposite policy. They design policies to estimate the usefulness of prefetches and control scheduling priorities.

Co-Design with Address Mapping and Prefetcher

The Minimalist Open-Page policy of Kaseridis et al. [48] tries to keep a row open long enough to exploit a minimal number of row buffer hits, while avoiding the pitfalls of keeping the row open for too long (unfairness). They first introduce an address mapping where four consecutive cache lines are placed in one bank, then the next four are placed in a different bank, and so on. For a single thread, this provides a good balance between row buffer locality and bank-level parallelism; for a group of threads, it automatically ensures that one thread will not monopolize the scheduler with a long burst of row buffer hits. Kaseridis et al. also observe that most row buffer hits are the result of prefetches generated by a simple stride prefetcher. Therefore, a row is precharged immediately following a prefetch-dictated burst of accesses. The scheduler priorities are determined by wait time (age), prefetch vs. demand fetch, prefetch distance, and the extent of memory-level parallelism (MLP) in the thread. This co-design with the prefetcher and the address mapping policy results in a simple scheduler.

Co-Design for Multiple Memory Controllers

High-performance processors typically have about a handful of memory channels that are independently controlled by their respective memory controllers. The ATLAS memory controller of Kim et al. [49] introduces co-ordination among these memory channels to identify threads that may be receiving lower levels of service. To reduce overheads and improve scalability, this co-ordination is performed at the end of a relatively large time quantum. The thread priorities in the next quantum are the inverse of the service levels received in the previous quantum. Awasthi et al. [50] note that pages can also be placed and migrated such that the overall burden on multiple memory controllers is reduced. When a new page is created, it is assigned to a memory controller based on a cost function that factors in distance from the requesting core, the load at that memory controller, and the average row buffer hit rate at that memory controller. If there is significant load imbalance among memory controllers, pages are moved to equalize the load. Note that the minimalist open-page policy [48] scatters a page across many banks and this can be easily extended to scatter a page across many memory channels as well. This may automatically help equalize load across memory channels, although, it may not promote page affinity (short distance between requesting core and data) in a NUMA system.

4.4 DISCUSSION

Much of the low-hanging fruit for intelligent memory schedulers has already been picked. There is likely little room for improvement, even for multi-core workloads with varying memory demands. Emerging memory systems like HMC and HBM even offer many more banks and higher bandwidth, potentially reducing the impact of scheduling tricks. Co-design with other components may have more potential. For example, certain code optimizations or accelerator configurations may be more amenable to efficient memory scheduling.

CHAPTER 5

Data Placement

Many factors impact the performance and energy efficiency of off-chip DRAM systems—DRAM timing parameters, parallelism, and row buffer locality. Therefore, smart data placement within DRAM, that in turn impacts timing and locality, is important and is covered in Section 5.1. Data compression deserves its own discussion in Section 5.2.

5.1 DATA INTERLEAVING

In Section 2.7, we discussed how bits of the memory address can be interpreted in different ways. The address mapping policy determines how data blocks get placed in different banks, ranks, and channels. Conventional wisdom says that applications with high spatial locality should map consecutive cache lines to the same bank so they yield a large number of row buffer hits. While this is certainly helpful from an energy perspective, latency can be lower if (say) four consecutive cache lines are fetched in parallel from four channels than if they were fetched sequentially from a single row buffer. In the past decade, we've therefore seen a gradual shift in this conventional wisdom.

Kaseridis et al. [48] observe that the locality in most pages can be exploited by placing four consecutive lines in a single row buffer; the other sets of four cache lines in a page are scattered across the banks, ranks, and channels to maximize parallelism. They claim this is a sweet spot, especially when combined with a prefetcher. Along similar lines, Sudan et al. [51] propose OS involvement in co-locating portions of jointly accessed pages in the same row buffer. A couple of works [52, 53] have also observed that for some workloads with specific strided access patterns, scatter-gather mechanisms can help, e.g., using custom data placement and on-DIMM controllers to issue different CAS signals to each chip in a rank [52].

When scattering requests across all available banks, a commonly employed trick is to generate the bank id by XOR-ing the bank address bits with higher-order bits in the address [54]. Other works have also highlighted the performance degradation because of bank conflicts; the Duplicon Cache creates copies of blocks so the memory controller has conflict-avoiding options when fetching a block [55]. Other factors to consider when placing pages in a multi-channel system are the load on each channel and the on-chip network latencies between cores and memory controllers [50], or the interactions with a low-latency (say) HBM that can serve as addressable memory or cache [56]. Bojnordi et al. [44] also show performance and energy benefits with a memory controller that can adapt the address mapping policy per workload.

We have thus seen a shift toward prioritizing parallelism instead of row buffer locality. A catalyst for this shift is the observation that multi-core workloads are much more likely to yield row buffer misses than row buffer hits. Memory systems for multi-core servers use adaptive (often proprietary) scheduling policies that track workload behavior to close a page more or less aggressively. To save activation energy, there will also be a gradual move toward narrower row buffers [57]. New devices like the HMC employ relatively narrow 256-byte row buffers [33] and instead derive high performance with high parallelism across hundreds of banks.

5.2   MEMORY COMPRESSION

The volume of data generated daily continues to expand. Compression will be vital in taming the challenges posed by big data. In fact, the highly popular HBO sitcom Silicon Valley has a fictional compression technology at the center of its high jinks.

Compression is frequently employed in persistent storage systems. At lower levels of the memory hierarchy, density is much more critical than latency. Therefore, it is worthwhile to spend thousands of cycles on compression/decompression and book-keeping if it can cut storage needs by 2×. However, compression is typically not employed in DRAM-based main memory systems. Main memory accesses cost only a few hundreds of cycles; the latency overheads of compression/decompression may therefore not be tolerable. The main memory and the page table are also central to the operating system; modifying the paging mechanisms to integrate compression would therefore require a significant overhaul of the OS.

In recent years, several breakthroughs have been made that may allow compression to be meaningfully integrated into the main memory system. We'll first discuss some of the early work in memory compression and the challenges they faced. We'll then discuss how some of these challenges have been addressed recently. The key takeaway is that compression is a lot more palatable if its features are diluted.

5.2.1   IBM MXT

IBM introduced Memory Expansion Technology [58, 59] nearly two decades ago to boost available memory capacity. For a number of relevant workloads, they show a compression ratio (aka compressibility) of over 2×, with a few workloads achieving compressibility of about 6×. They employ the lossless LZ77 compression algorithm [60]. This algorithm achieves high compressibility while requiring a compression/decompression latency of 1 byte/cycle. To mitigate this high latency, the data being compressed is split into smaller blocks and each is compressed in parallel.

In MXT, the L3 cache deals with 1 KB blocks. This is of course a much larger granularity than what is typical in today's bandwidth-starved and power-hungry servers. After compression, a 1 KB block is placed in 0–4 256-byte sectors in physical memory. A 16-byte metadata entry, also in physical memory, stores pointers to these sectors. For a read, the metadata must be first read from memory; after parsing the metadata, multiple sectors may have to be read from memory; these sectors are then decompressed in parallel to yield the final uncompressed block. Even with the implemented parallelism, the latency for compression and decompression can exceed a hundred cycles given the large block size. Thus, random memory accesses are quite expensive, partially because of the large block size, partially because the metadata access is on the critical path, and partially because of long compression/decompression latency.

5.2.2   EKMAN AND STENSTROM

Large granularity memory accesses are not practical when bandwidth and power are at a premium. In their 2005 paper, Ekman and Stenstrom [61] designed a compressed memory implementation that is more in line with modern DDR3/DDR4 hierarchies, i.e., caches and memories deal with the more standard 64-byte blocks. The IBM MXT fetched and decompressed an entire 1 KB block to locate individual words in the block. But when fetching blocks at a finer granularity, mechanisms are required to quickly locate the specific block being requested within a larger compressed page. While this is a non-issue for traditional uncompressed memory systems (the block's physical address pinpoints its location), a compressed memory system no longer places a block in a pre-designated location. Since all the compressed blocks are packed together to reduce the size of a page, the starting address of a block depends on the compressibility of prior blocks in that page.

To address this problem, Ekman and Stenstrom [61] add metadata to every page table entry to track the size of each compressed block. This information is stored on chip in TLB-like structures. To estimate the starting address of a block, additional logic is required to add the sizes of all preceding blocks in that page.

This approach of packing compressed blocks in a page leads to other complications. If a block is modified, its newly compressed version may be larger than its original compressed version. To make room for this block, all the subsequent blocks in the page have to be moved. On the other hand, if the newly compressed version is smaller, the subsequent blocks have to be moved to fill in the hole. To alleviate the overheads from this jockeying for space on every block modification, Ekman and Stenstrom introduce a few features, e.g., allowing only a few compressed block sizes and introducing sub-pages. They also use simple low-latency compression algorithms that rely on an abundance of zeroes in the data block [62, 63]. The paper thus explores a design point that offers medium compressibility and fine-grained accesses, but more metadata/OS/logic complexity in managing individual blocks.

5.2.3   LINEARLY COMPRESSED PAGES (LCP)

The work of Pekhimenko et al. [64] continues this line of innovation and further reduces some of the complexity overheads. First, the authors introduce a low-complexity compression algorithm, *Base Delta Immediate* or BΔI, that achieves a sufficiently high compression ratio in the neighborhood of 1.5$\times$ [65]. BΔI is based on the observation that most words in a block are in close proximity to zero or to a base word in that block. Compression is therefore achieved by tracking the base word, and offsets of other words from this base word or from zero. Additional metadata in the compressed block indicates the sizes of the base word and offsets. BΔI has quickly become the gold standard for memory compression projects.

Next, to reduce the complexity in locating compressed blocks, Pekhimenko et al. [64] introduce a novel *linearly compressed page* (LCP) layout for compressed pages. All the blocks in a compressed page are required to have a fixed (compressed) size—this makes it trivial to find the starting address of any compressed block. Of course, some blocks in a page will not meet this requirement—those blocks are stored uncompressed in an exception region at the end of the page. To manage the exception region, additional metadata (that fits in a cache block) is required. This metadata stores bit vectors and pointers to quickly locate blocks in the exception region. Since the metadata for a page is vital for efficiency and correctness, the authors introduce a metadata cache—this is especially effective when an application exhibits spatial locality within a page. Further, a few bits of metadata must be included in page table entries and TLBs.

With the LCP approach, most compressed blocks can expand or shrink when modified without requiring other blocks to be shuffled (in some cases, if a compressed page exceeds its allotted capacity, the entire page must be copied into a larger region). Pekhimenko et al. also allow a single memory access to fetch two compressed blocks if they fit, thus improving bandwidth utilization. Overall, LCP represents an attractive design point that uses compression to significantly grow memory capacity, while alleviating most metadata management overheads. Very recent work by Choukse et al. [66] builds upon an LCP-like architecture; it characterizes the sources of metadata access inefficiency and reduces them with a number of techniques: cache line alignment, predicting and preparing for page re-organization because of overflow and underflow.

5.2.4   NEARLY OVERHEAD-FREE MEMORY COMPRESSION

Clearly, the higher memory capacity enabled by compression is its raison d'être. But the resulting variable page sizes, metadata management, and blocks jockeying for space are also the primary barriers to widespread commercial adoption.

Some recent works [67–69] try to lower this barrier by eschewing the capacity benefits of compression. They claim that compression already offers several other benefits: (i) higher effective bandwidth, parallelism, and energy-efficiency if the memory controller can fetch smaller blocks; and (ii) higher reliability and energy efficiency with coding techniques that exploit the spare space created by compression.

These works do not change the starting address of a block, relative to an uncompressed memory system. So they do not enable more blocks in the memory system, i.e., no increase in memory capacity. But since each compressed block occupies less space than in the baseline, the trailing bytes in such blocks are empty. These trailing bytes can be used to store stronger ECC codes or DBI (data bus inversion) codes that invert words in a manner that reduces data transmission energy [67]. Or, if the memory interface allows it, the trailing bytes need not be fetched, thus lowering energy and the pressure on the memory bus. Sathish et al. [68] exploit the fine-grained memory accesses in GDDR5 to enable this.

Shafiee et al. [67] introduce an implementation, MemZip, that achieves a similar effect with commodity DDR3/DDR4 memory devices. To enable fine-grained memory accesses, they assume a memory interface that supports rank subsetting. A conventional DDR3/DDR4 rank has a 64-bit interface (without ECC) and a minimum burst size of 8, requiring 64-byte reads/writes. Rank subsetting, as explained in Section 6.2, uses mini-ranks with fewer chips and narrower interfaces, thus allowing reads/writes at finer granularities (lower energy). Different blocks can also be fetched simultaneously from different mini-ranks over a single shared 64-bit bus, thus boosting effective bandwidth as well. Shafiee et al. explore different mini-rank organizations that achieve varying trade-offs in terms of latency, parallelism, and energy. Since BΔI can achieve an average compression ratio of 1.5, MemZip can boost effective bandwidth by 1.5$\times$, yielding a significant performance boost for bandwidth-intensive systems, even if they aren't constrained by memory capacity.

The MemZip approach offers other key advantages. It is trivial to locate a block—the starting address is the same as in the baseline uncompressed memory. Because a block is free to expand and shrink in the 64 bytes allocated to it, neighboring blocks are unaffected and need not be shuffled when a block is modified. Each block needs a single bit of metadata to indicate if it's compressed or not. If the block is compressed, the leading bytes in the block can indicate the block size and any other metadata information required for decompression. This single compressibility bit per block can be stored as part of the page table/TLB or in a separate metadata cache. The small size of this metadata per block implies higher cacheability than LCP's metadata cache. Since MemZip does not impact page size or memory capacity, the impact on the OS is minimal (apart from the need for metadata storage). The key drawback of MemZip is its reliance on rank subsetting—an interface that is not currently part of the DDR standard, but appears to be on the roadmap for DDR5.

More recent work by Deb et al. [69] builds on the MemZip design to eliminate OS involvement in memory compression. This is done by integrating the compression metadata with the word that stores ECC information.[1] The authors show that Hamming Codes can be modified to provide similar SECDED coverage as the baseline, while sparing a single bit per block to track if the block is compressed or not. Compression and decompression can therefore be handled entirely in memory controller hardware along with the logic that handles ECC, without requiring the OS to reserve/manage a region of physical memory for metadata storage. The downside of this approach is that when performing a read, the size of the compressed block is only known after part of the block has been read. If the block size is known beforehand, the memory controller can engage the bare minimum number of chips (a mini-rank) required to fetch a compressed block, thus improving performance and energy. Deb et al. therefore intro-

---

[1]For analysis of another technique that cleverly combines ECC and compressibility information, see the discussion of COP [70] in Section 8.2.

duce a simple predictor (based on PC or page id) with 97% accuracy to estimate the size of the compressed block before the read operation is initiated. The authors also explore different mapping policies that engage different mini-ranks for different blocks, thus promoting uniform activity across memory chips and higher parallelism. A more recent paper, Attache [71], also addresses similar problems. Unlike the approach of Deb et al. that needs ECC DIMMs, Attache uses a golden 15-bit keyword at the start of the block to indicate if it is compressed. If an uncompressed block starts with the same keyword, another bit indicates if the block is uncompressed; the true value of that bit has to be fetched from another region of memory. Thus, additional metadata accesses are required only in the rare case where an uncompressed block happens to start with the golden keyword. Attache also relies on mini-ranks and introduces a three-tier predictor to decide how much of the block to fetch.

The end product of these recent innovations is *compression-lite*: a subset of compression benefits with zero software/OS overheads and a small overhead in memory controller hardware. These innovations only apply to non-commodity memory interfaces (which may be common in the near future). And, notably, these innovations do not yield an increase in effective memory capacity.

5.2.5   PTMC

Young et al. [72] introduce Practical and Transparent Memory Compression (PTMC) to address the weaknesses described in Section 5.2.4, namely the reliance on mini-ranks. They too focus on the bandwidth benefits of compression, not capacity benefits. Without mini-ranks, a block fetch from memory is always 64 bytes wide and may either bring in an uncompressed block P at that address, or it may bring in 2 or 4 consecutive compressed blocks starting at that address. To avoid the need for separate metadata, a compressed block includes a 4-byte golden keyword to indicate that it's compressed. Uncompressed blocks that include the golden keyword are stored in inverted form and tracked in a small on-chip table. Such embedded metadata also makes this approach compatible with non-ECC DIMMs. Further, to identify if a block has been compressed along with prior lines, a line location predictor is required. The authors show that such a predictor has a high accuracy of 98%. The authors also point out that their approach is similar to that used by the Qualcomm Centriq 2400 processor [73], which also uses compression to reduce bandwidth and pack a 128-byte block into a half-block when possible. The Centriq system relies on ECC DIMMs for metadata tracking.

5.2.6   ACTIVE MEMORY EXPANSION IN THE IBM POWER PROCESSORS

To increase effective memory capacity, significant OS involvement is inevitable. IBM has clearly led the charge in the effort to maximize memory capacity, first with the MXT approach (circa 2001) described in Section 5.2.1, and more recently with Active Memory Expansion (AME) [74] in Power7 and Power8 systems.

All the approaches described in this section compress the entire memory system and require decompression for every memory access—AME is notably different. It splits the memory address space into two pools—a compressed pool and an uncompressed pool. Recently accessed pages are placed in the uncompressed pool and accessed at low latency. Pages that have not been accessed recently are compressed and moved into the compressed pool. This takes the overheads of compression and metadata processing off the critical path, especially when applications exhibit high spatial and temporal locality. To some extent, the uncompressed memory behaves like a conventional memory system, and the compressed memory behaves like a low-latency swap space. In contrast to MemZip and PTMC, AME targets an increase in memory capacity and does not attempt to improve bandwidth/parallelism/reliability. AME allows the user to define the desired memory expansion factor or compressibility. The OS then automatically defines the sizes of the two pools to achieve that expansion factor. The compression and decompression are performed on the CPU, increasing CPU utilization by under 10%. Newer Power processors have introduced custom logic to perform the compression and decompression [75].

5.3 DISCUSSION

Smart data placement in memory continues to be an intriguing problem with the potential for high impact. While many-banked memories (HBM, HMC) with narrow row buffers favor policies that spread data, there are other considerations for hardware/OS data placement policies. For example, multiple efforts, discussed in Section 6.2, try to introduce variable latency within DRAM. Therefore, identification of hot pages and load balancing will continue to be important.

Compression is an area where we'll likely continue to see innovation, simply because it is only just garnering attention from industry. The breakthroughs in the last five years will likely have commercial impact, e.g., the use of compression to reduce bandwidth in Qualcomm's Centriq system [73]. These breakthroughs have shown how compression, or restricted/diluted versions of compression, can largely be transparent to the OS, and be compatible with DDR standards or with future standards like DDR5 that allow finer-granularity access. Industry is clearly paying close attention to reliability and security. It seems natural that compression be included in such ecosystems. For example, as we'll discuss in depth in Chapter 11, Intel SGX implements a region of secure memory accessible at block granularity and a region of secure memory accessible at page granularity. This is very reminiscent of how IBM AME manages compressed data. So one could imagine future systems where a page is decrypted, uncompressed, and error-corrected before it is placed in a processor-accessible region of memory.

CHAPTER 6

Memory Chip Microarchitectures

Swiss watches are engineering marvels, and so are DRAM chips. Swiss watch makers would be justifiably upset if you took a look under the hood, decided to poke and prod, and suggest alternative designs. You'd receive the stern "You think we didn't consider that?" look. A DRAM designer's look would be even more stern: "We operate at very tight margins and you think we didn't consider that?!"

For years, architects had left DRAM microarchitecture to DRAM manufacturers; most ISCA regulars (myself included) didn't know if a row buffer was an array of latches (it isn't). As concerns over datacenter operational cost and DRAM power grew, a few groups (mine included) started questioning DRAM microarchitecture [19, 76]. Our talk at ISCA [19] raised a few skeptical eyebrows; DRAM engineers scoffed at our area analysis; subsequent papers schooled us on how DRAMs were actually designed; follow-up papers then schooled the subsequent papers on how DRAMs were *actually* designed. To this day, DRAM microarchitecture is shrouded in mystery. We understand the overall nuts and bolts, but it is very difficult to re-create an industry-caliber prototype design in an academic lab, poke at it, and credibly quantify the impact on area, cost, cycle time, etc. Given what we know today about DRAM microarchitecture, I am skeptical that the area analysis in our original paper is in the right ball-park for a commercial DRAM chip; but that basic idea of fine-grained row activation [19, 76] has persisted [57, 77–79], and has been applied in commercial settings in HMC, HBM, and DDR4.

In this chapter, we'll first discuss the basic organization of a DRAM chip [77–80], followed by a tour of recent ideas to improve upon this organization.

6.1 BASICS OF DRAM CHIP MICROARCHITECTURE

The DRAM chip has a very regular and hierarchical organization. A chip is organized into 8 (DDR3) or 16 (DDR4) or more (HBM/HMC) banks. The banks typically share a bus that moves data between the banks and the IO pins. A bank is partitioned into multiple subarrays (a row of mats shown in Figure 6.1), and each subarray is partitioned into a number of mats. A mat is typically 512 rows and 512 columns of DRAM cells.

When a DRAM block is being read, the address is sent with two commands: the Activate or RAS (row address strobe) command that identifies a row, and the Column-Read or CAS (column address strobe) command that identifies a specific block in that row. The RAS

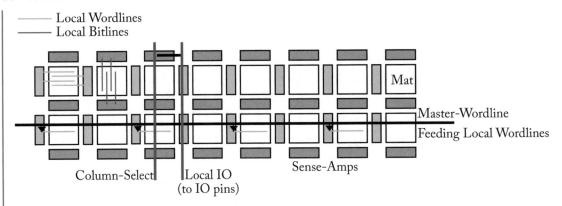

Figure 6.1: Two rows of mats in a bank are shown here. Each row is called a subarray. The sense-amplifiers are shown in orange, while the row decoders are shown in light blue. Other major interconnects are shown in green (local wordlines, implemented in poly), red (local bitlines, bottom metal layer), thick black (master wordline and bus feeding local IO, middle metal layer), and thick blue (column-select and local IO, top metal layer). Note that local wordlines and local bitlines are staggered.

bits are first used to activate one of the subarrays in a bank. The RAS bits are also decoded to identify a row in that subarray; a master wordline then enables local wordlines (shown in green in Figure 6.1) in each mat of that subarray. This connects the cells in that row to their respective bitlines, shown in red. Each bitline is connected to a sense-amplifier that not only identifies the presence/absence of charge, but also restores charge back to the corresponding cell (since the DRAM read is destructive). At this point, the contents of a wide row have been captured in a few thousand sense amplifiers (shown in orange), which serve as the row buffer.

Next, the CAS bits are decoded and some of the sense-amplifier values are selected with column-select lines. A horizontal bus moves the selected values to the edge of the mat. From here, the values are moved over local-IO wires to the IO pins (see Figure 6.1). If the subarray has 8 mats and each mat has 4 local-IO wires, 32 bits are sent to the IO pins in a single cycle. Note that the mat and its local-IO may operate at a slow clock speed of say 200 MHz. These 32 bits may then be placed on 4 output pins (a ×4 chip), with a burst size of 8. If the output pins operate at 800 MHz and DDR, the bandwidth of the output pins matches the bandwidth of the bank's local-IO wires.

DRAM chips can use a number of techniques to increase density. DRAM chip microarchitectures also vary depending on which of these techniques are being employed. For example, a decoder may be shared by its right and left mats [78], and a sense-amplifier may be shared by the subarray above and below. Thus, every odd wordline (bitline) may be driven from the left (top), while every even wordline (bitline) may be driven from the right (bottom). This staggered organization, shown in Figure 6.1, allows the corresponding peripheral circuit (sense-amp or decoder) to span two columns or two rows, resulting in a denser layout.

The low-cost DRAM process only uses three metal layers [79]. Local wordlines are implemented with silicided polysilicon (shown in green in Figure 6.1). The bottom metal layer with the smallest wire width (pitch) is used for the vertical bitlines in each mat, shown in red. The next metal layer implements the wider horizontal wires (the master wordlines and the bus between sense-amps and local-IO, shown in thick black in Figure 6.1). The top metal layer implements the wider vertical wires that travel longer distances (the local-IO and the column-select lines, shown in thick blue).

6.2   DRAM CHIP INNOVATIONS

We next discuss some of the primary academic ideas to improve upon the basic DRAM microarchitecture. The effectiveness and applicability of some of these ideas vary based on microarchitecture assumptions.

Fine-Grained Activation

A sequence of papers [17–19, 57, 76–79] has observed that in conventional DRAM chips, the Activate/Precharge operation is one of the most expensive in terms of energy—about a quarter to half of overall DRAM energy on average and a much higher contribution with non-existant spatial locality. This is why DRAM timing parameters impose constraints on the number of Activates that can be issued in a given time window (*tFAW*). An Activate enables a long wordline in several DRAM chips; the contents of thousands of cells is sensed and restored by sense-amplifiers. Historically, this was viewed as a feature because this prefetch can lead to several lower-latency row buffer hits. In more recent times, multi-core workload characterizations [17, 19, 48, 51] have shown that the row buffer locality is limited and the overfetch to row buffers is mostly wasted energy. This has prompted a re-think of the DRAM microarchitecture—can we somehow enable a shorter wordline while minimally impacting the area of the peripheral circuitry?

Initial efforts [17, 18] kept the DRAM chip intact, but shrunk the width of the rank. Activation energy is reduced by using fewer DRAM chips per request. A given cache line transfer therefore has fewer DRAM pins and fewer memory channel wires at its disposal. The system essentially uses more narrow ranks, which reduces wait time, but increases the data transfer time. For most workloads, this ends up being better in terms of performance and energy. The downsides are that it deviates from the DDR standard, and makes error correction more complicated. For example, if the cache line is fetched from 4 ×8 chips, adding a fifth chip for error codes results in a significant 25% overhead—turns out that DDR5 is headed in this direction anyway because of escalating error rates.

Subsequent research focused on modifying the DRAM chip to reduce activation energy. Concurrent work by Udipi et al. [19] and Cooper-Balis and Jacob [76] takes advantage of the hierarchical nature of the wordline: each local wordline is AND-ed with a select signal, thus reducing the number of activated mats. The select signal is derived from the CAS bits, therefore requiring that the CAS command be sent early. Subsequent work [77] has added more DRAM microarchitecture detail to the design and shown that activating fewer mats has the negative side effect of reducing the local IO wires feeding the output pins. O'Connor et al. [79] overcome that problem for an HBM interface by also re-organizing the external channel, which we discuss shortly. Udipi et al. also advocate that an entire cache line be placed in a single mat on a single chip—this is a design point that favors low energy and high parallelism over latency and cost. Just like earlier proposals on sub-ranking [17, 18], this requires changes to the DDR standard. The authors also introduce in-DRAM checksums and RAID-like ECC so that cache lines can be re-constructed upon failure.

Zhang et al. [77] note that early attempts at fine-grained activation ignore the impact on local IO wiring. Conventional microarchitectures have horizontal and vertical wiring to move a few bits from each mat to the output pins. If only some mats in a subarray are activated, it either reduces the bandwidth fed to the output pins, or more horizontal and vertical wiring is required per mat. To reduce this overhead, the authors propose a Half-DRAM design, i.e., a not-too-aggressive fine-grained activation that cuts activation energy in half. In the Half-DRAM design, a decoder drives half a local wordline in its left and right mats. This allows half a wordline to be activated in every mat in a subarray. Since every mat is being activated, the local IO wires of every mat can be engaged, thus having zero impact on bank bandwidth. The lower power per activation also allows more simultaneous activations, thus improving performance. Ha et al. [78] analyze a similar fine-grained activation approach for a baseline that has the staggered wordlines and bitlines described earlier.

The work of Chatterjee et al. [57, 79] is the first to apply these approaches to HBM. The authors first segment the wordline to activate a smaller set of mats; the channel is also split into subchannels [57]. They then carry out a design space exploration [79] to understand the approach that most effectively improves HBM bandwidth by 4× over the state-of-the-art. They observe that of the 4 pJ/bit energy for HBM access, 1.2 pJ/bit is in activation, and about 2 pJ/bit is in planar data movement from the sense-amps to TSVs, and from TSVs to IO bumps. Since interposer interconnects are more energy-efficient than the on-die interconnects, it is better to scatter the IO bumps so that on-die interconnect lengths can be reduced. The authors combine this approach with fine-grained activation to reduce both activation energy and on-die interconnect energy. Given the low bandwidth of a fine-grained bank, it is connected to a narrower channel with its IO bumps located nearby. This increases the data transfer time for a cache block, but supporting many narrow channels is a worthwhile trade-off for GPU/HBM systems that are less sensitive to high latency and where the workload has high memory-level parallelism. O'Connor et al. acknowledge that getting an entire cache line from fine-granularity DRAM makes error-correction more problematic, requiring wider mats and/or faster channels [79].

Lee et al. [81] combine a fine-grained activation approach with a technique that treats writes differently. This is motivated by writes exhibiting much lower row locality, and modifying only a few words in a cache line. Reads are handled the same as in the baseline, but writes enable local wordlines in only some mats. To reduce area overheads, the local IOs in those mats are not augmented, i.e., the overall bank is forced to operate at lower bandwidth. To alleviate the impact on performance, only a subset of the words in the cache line—the words that are dirty— are written into DRAM. The last level cache must therefore track dirty words in a line and send a mask to the DRAM chips. The DRAM chip then performs a partial row activation and only writes dirty words to its corresponding mats.

Sub-Array Level Parallelism [82]

Unlike the efforts above that try to reduce the energy and size of an Activate command, Kim et al. [82] exploit the hierarchial nature of a DRAM chip to increase parallelism and improve performance. A bank has multiple subarrays, with some circuits independent and some shared. The authors observe that with no changes to the DRAM chip, an Activate in one subarray can be performed while a precharge is being performed in another subarray. Conventional memory controllers do not allow this because the subarrays within a bank are not exposed to the memory controller; the memory controller therefore makes the worst-case assumption that the precharge and activate are in the same subarray and therefore serializes the two operations. Kim et al. augment this technique by further allowing an activate in one subarray while another subarray is still performing its write recovery. This requires a minor change where a subarray must latch its row address so that the next subarray can receive a new row address. Finally, the authors allow each subarray to keep its row open; the memory controller uses a new command and each subarray maintains a bit to keep track of one subarray that is deemed "current" and whose sense-amplifiers are connected to local IO. The other subarrays stay in the background with valid row buffers in case they can yield row buffer hits by turning into the current subarray. Thus, with small changes to the DRAM chip, each subarray can be made more independent, and enabling intra-bank parallelism.

Variable Latency DRAM

A few works have modified the DRAM chip microarchitecture to offer variable latencies. In Section 9.2.1, we discuss variable latency techniques based on the charge level in a row. Here, we focus on designs where some rows always offer lower latency than others.

Lee et al. [83] observe that a significant contributor to high DRAM latency is the long bitline in a mat. They therefore introduce an isolation transistor in every bitline; depending on the state of the isolation transistor, the sense-amplifier is either connected to a short bitline or a much longer bitline. In other words, the isolation transistor partitions a mat into two tiers; rows in the tier closer to the sense-amplifier exhibit faster timing parameters than rows in the further tier. The primary overhead in this approach is the area penalty of implementing and controlling the isolation transistors, and the OS policies that can improve performance by placing hot pages in the faster tier.

While the work of Lee et al. introduced non-uniformity within every bank, Son et al. [84] introduce a design with uniformity within each bank, but non-uniformity among banks. They observe that banks closer to the central IO pins can exhibit lower latencies because of the shorter global wires that must be traversed. To further improve the speed of these central banks, their mats are implemented with shorter bitline lengths, i.e., a mat has wider rows and fewer columns. Meanwhile, Shevgoor et al. [85] also observe that the location of a bank in a 3D stack, relative to the power delivery network, can impact the timing parameters for that bank.

For all the above variable-latency DRAM chip organizations, the migration of hot pages from slow to fast regions is important. RowClone [86] shows how copies can be performed between rows in a single subarray, while two other works introduce new circuits to efficiently perform these copies between different banks [87, 88].

Ro et al. [89] recognize the difficulty in managing variable-latency regions, so they exploit some of the techniques mentioned above to create a uniformly lower-latency DRAM chip. Shorter bitlines are employed in banks that are further from the IO pins, thus offsetting their disadvantaged location. The data transfer from a bank to the IO pins can also be pipelined to accommodate varying wire lengths within the bank.

Another body of work [90–93] uses profiling of DRAM chips or DRAM chip regions to estimate latency as a function of temperature and parameter variation. Chandrasekar et al. [91] propose aggressive timing parameters for a rank that shave the generous timing margins inherent in DRAM chips. Lee et al. [90] adapt the timing parameters for the rank as a function of temperature. Zhang et al. [92] and Taassori et al. [93] exploit intra-chip parameter variations and re-organize how banks are ganged together in a rank to lower latency.

6.3   DISCUSSION

Research on innovative DRAM chip microarchitecture certainly contributes to the marketplace of ideas, but there are some inherent risks. There is inevitably some fumbling in the dark and results must be taken with a grain of salt, more so than other simulation-based architecture results. Working with industry teams can help alleviate some of the risk. While fine-grained activation appears to be commercially attractive because of its low energy and higher parallelism, the idea of variable latency within a DRAM chip has made no inroads. Given how optimized DRAM circuits are, and given the tight margins for the commodity DRAM industry, the room for improvement may be relatively small for DDR chips, and potentially higher for emerging memory devices that favor performance or energy over cost.

CHAPTER 7

Memory Channels

It is well known that modern systems are increasingly being limited by the cost of data movement [94]. The data movement between DRAM and the processor is one of the largest contributors to this emerging bottleneck because it involves off-chip links.

A variety of interconnects are used to connect memory devices to the processor. The choice of interconnect influences the memory bandwidth, latency/energy for data transfers, and the memory capacity that can be supported. These interconnects can contribute nearly half the total power consumed within the memory system [2, 95].

Earlier, in Chapter 3, we discussed a number of commercial memory products. That was an initial exposure to how memory devices are connected to the processor. In this chapter, we'll first go more in-depth into various interconnect options, followed by a discussion of recently proposed improvements to memory channel architectures.

7.1 THE BASICS OF PARALLEL AND SERIAL INTERCONNECTS

At a high level, there are two major types of interconnects that connect memory devices—parallel and serial.

7.1.1 PARALLEL BUSES

DDR buses are parallel interconnects. They use single-ended signaling, i.e., the driver sets the voltage on a single wire to high and low, and this is detected by the receiver. Ideally, when the transmitter raises the voltage to high, the receiver should observe a nearly identical voltage curve. But depending on the load on the bus, noise, interference, reflections, etc., the voltage curve at the receiver is slightly distorted. The quality of this voltage curve at the receiver is referred to as signal integrity. An "eye diagram" helps measure the signal integrity by tracking the voltage curve for a 0 to 1 transition and a 1 to 0 transition. If the curve looks like an "open eye," then it's easy for the receiver to make the distinction between 0 s and 1 s. This is measured by looking at the aspect ratio of a rectangle that fits within the eye.

As the name suggests, a parallel interface is implemented with a number of parallel single-ended wires. The cross-coupling interference among these wires further degrades signal integrity. Some of these wires carry the clock signal so the transmitter and receiver agree on clock edges. At higher bus frequencies, the skew among the data and clock wires increases, as does the interference among wires. To preserve high signal integrity, DDR3 can operate at a maximum frequency of just over 1 GHz, and DDR4 can operate at a maximum frequency of about 1.6 GHz (with data being transmitted on both rising and falling edges). The maximum frequency is lower if multiple DIMMs are plugged into the channel. The energy per bit transfer in a DDR bus is significant because of the full voltage swing. The energy for DDR3, DDR4, LPDDR2, and GDDR5 technologies that rely on on-board interconnects is 70 pJ/bit, 50 pJ/bit, 40 pJ/bit, and 14 pJ/bit, respectively, while that of HBM2 that relies on parallel interposer interconnects is 3.9 pJ/bit [79, 96, 97].

7.1.2 SERIAL BUSES

A serial bus is increasingly more popular as we move to high-frequency regimes. A serial link can offer high bandwidth per pin and low energy per bit. But as we'll describe next, it also has a few disadvantages. Note that a serial link is also referred to as a SerDes link because a wide packet has to be sent serially (one bit at a time) over the link and has to be put together (deserialized) at the receiver.

Communication on a serial link is performed with low-voltage differential signaling (LVDS) on a pair of wires. While this may seem like a high pin overhead (two pins to send one bit), that overhead is more than recovered with the speed and energy efficiency that LVDS enables. The energy efficiency is enabled by the low voltage swing on the wires. The low voltage swing also leads to low cross-talk, thus playing a role in enabling higher clock speeds. The potential skew among the many wires at high frequency is managed by making each pair of wires independent and embedding the clock in the data being transmitted by that pair of wires. How is the clock embedded in data? This is achieved with data encoding techniques that guarantee roughly equal numbers of 1 s and 0 s, and frequent transitions between 1 s and 0 s in every packet. An example popular encoding is 8b/10b, where every 8b packet of data is encoded into a new 10b packet of data. However, the link must continuously carry 10b data packets to keep the transmitter and receiver synchronized, even when no data communication is required. In other words, these dummy data transfers in a serial link consume energy while a corresponding parallel bus consumes very little energy when idling.

Why are serial links attractive? A single serial link (with two wires/pins) can operate at frequencies of tens of giga-hertz (since skew is not a concern). Thus, it takes very few wires/pins to achieve the target bandwidth of a system even after factoring the 25% bandwidth overhead imposed by 8b/10b encoding. Thanks to the low swing nature of the signaling, the energy per bit is also lower than that of a parallel interface.

So why isn't every interface serial? The energy per bit advantage can disappear if the link utilization falls below a certain threshold. This is because a serial link is always transferring 10 bit packets while a parallel link can idle efficiently. This drawback may be overcome as the power-up and power-down modes in serial links are made more efficient [95, 98]. Serial links also tend to be short. At longer distances, or when load increases, the energy and frequency advantages start to erode. Therefore, a number of short point-to-point serial links may be required to create a large-capacity memory system. Additionally, a serial link requires a non-trivial amount of logic at either end to perform 8b/10b encoding/decoding and to packetize the data. Placing this logic on every DRAM chip can introduce a significant area and cost overhead. Finally, as analyzed in a recent study [98], a serial interface can add extra latency for every hop and is less ideal for latency-sensitive workloads.

The Micron HMC's link is a prominent example of a serial interface. An HMC has 4 or 8 serial links that can be used to connect the HMC to processors and other HMCs. Each link is itself composed of 16 upstream lanes (each lane uses a pair of differential wires to carry a bit per cycle) and 16 downstream lanes. The link operates at frequencies of 10 GHz and higher. At 10 GHz, a single link offers 20 GB/s upstream and 20 GB/s downstream bandwidth. About half the HMC power is in the SerDes signaling circuits, in the neighborhood of about 5 pJ/bit and projected to go down to 1 pJ/bit [33, 99, 100].

The HMC is also a perfect platform for a serial link. The HMC makes a number of trade-off choices that increase bandwidth and lower latency, while incurring a penalty in cost. Each HMC package has 4–8 DRAM dies and a logic chip. The SerDes circuits can therefore be placed on a logic die instead of on a DRAM die; this allows the high-frequency SerDes circuits to use a higher-quality logic process; it also shelters the DRAM die from the demands of a serial interface, allowing reuse of existing high-density DRAM layouts. A single SerDes hop enables access to 4–8 DRAM dies, thus reducing the hops to reach a given memory capacity.

7.1.3   DISCUSSION

We have summarized the pros and cons of both parallel and serial links. To strike a balance, the two can also be combined, and this has even been done commercially. As introduced in Section 3.2, a BoB chip can be used as a bridge between different interconnects that are optimal for the processor and for memory chips. On one side of the BoB are serial links for high bandwidth/pin to the processor; on the other side of the BoB are parallel DDR links for a low-cost high-capacity memory system.

Our discussion above has focused on the more popular electrical interconnects currently in use. Other exotic technologies have also been proposed for processor-memory communication, most notably photonics. While photonic interconnects were heavily studied in the past decade, they don't seem to have caught on, especially as serial interfaces have provided high bandwidth at low cost. The key advantage of photonics is that it provides very high bandwidth that exceeds the capabilities of electrical pins (but perhaps not the capabilities of many CMOS chip metal layers), and low energy when traversing distances slightly beyond the size of a chip. As such, it was a perfect fit for off-chip memory connections. Two ISCA papers delved into the best interface to connect memory to the processor using photonics [101, 102]. The work of Udipi et al. [101] proposed an interface die in a 3D stacked package of memory dies that houses the optical interface, thus sheltering the DRAM dies from the complexities of "exotic" interconnects (similar to the logic die in an HMC).

As a separate issue, note that the BoB chip or the interface die in a 3D stacked package [101] can also take ownership of low-level memory details, i.e., the processor simply provides "Read AddressX" or "Write AddressY" commands, while low-level management of timing constraints and scheduling can be performed by logic on the BoB chip or interface die. This can help reduce the latency overheads of repeatedly using the serial or photonic interface.

In the coming years, we are likely to see a hierarchy of interconnects in the memory system. For example, a processor may be connected to an HBM cache with parallel links (the use of parallel links is a key distinguishing feature of HBM, compared to HMC); the processor may also use serial links to connect to a BoB, which then uses parallel links to DDR DIMMs. To assist with explorations of this potentially vast design space, CACTI 7 [2] adds models for both parallel and serial interconnects, with varying lengths, load, frequency, etc. The following figures from the CACTI 7 paper show some of the insight revealed by such design space explorations.

In Figure 7.1, we see two example memory architectures to implement a 4 GB HMC cache, backed up by 128 GB of longer-latency memory. The first approach (Figure 7.1a) uses two DDR4 channels and two quad-rank LRDIMMs per channel. The second approach (Figure 7.1b) uses a network of 32 HMCs. Figure 7.1c then shows a breakdown of memory power for either approach, with interconnect (I/O power) shown separately. An analysis like this reveals that a hybrid HMC/DDR4 architecture is best, and highlights the role of interconnects and data movement in future memory organizations. Figure 7.2 also shows how DDR3 and DDR4 interconnect power varies for different system parameters: DIMMs per channel (DPC), bus frequency, read-intensive or write-intensive workloads, and DIMM type.

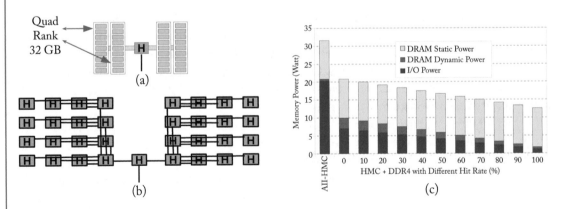

Figure 7.1: Memory organizations with (a) one HMC "cache" that is backed by 2 DDR4 channels and 4 quad rank LRDIMMs and (b) an HMC-only organization with one HMC "cache" backed by 32 other 4 GB HMCs. (c) Memory power for these two organizations, as a function of the hit rate in the HMC "cache." Data derived with and reported in CACTI 7 [2].

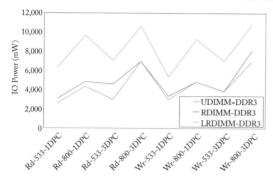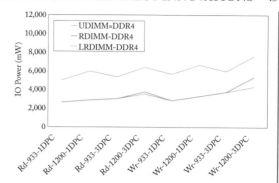

Figure 7.2: Memory I/O power (mW) for various DDR3/DDR4 design points. Each design point represents a different combination of DIMM type, read/write intensity, frequency, and DIMMs per channel (DPC). Figure appears in Balasubramonian et al. [2].

7.2   MEMORY INTERCONNECT INNOVATIONS

7.2.1   RECENT WORK

In this section, we'll first discuss recent attempts to improve the memory channel. Energy of parallel links can be reduced with better data encoding and frequency/voltage throttling; meanwhile, serial links can save energy by shutting off during idle times. Performance can be improved by matching the workload to the bandwidth supported by various links.

Alloy [98]

    The authors of the Alloy work make several observations about state-of-the-art serial links. They show that because of the encoding/decoding pipeline, serial links can add an extra 10-30 ns to latency. While some of this may not be on the critical path (a memory request can be encoded while it waits in the scheduling queue), a significant fraction is. On the other hand, they can offer 4× more bandwidth per pin than a parallel link. In some systems, the demands of applications are bimodal, e.g., in an integrated CPU/GPU system, the CPU demands a low-latency memory while the GPU demands a high-bandwidth memory. In such systems, the processor pins can be allocated to memory regions accessible via serial links or via parallel links. The OS then maps latency-sensitive (CPU) pages to the region accessible with parallel links and bandwidth-sensitive (GPU) pages to the region accessible with serial links. It is worth noting that in many systems, queuing delay is a significant fraction of memory latency; in such systems, a high-bandwidth serial interface may help reduce overall latency even if it introduces encoding/decoding delay. In addition to the above hybrid memory network, Alloy also introduces various power-down modes for the serial link. Their characterization shows that the active link consumes 63 mW/lane, while a lane in standby consumes 13 mW (mostly leakage) and can be woken up in as little as 10 ns. The hibernate mode consumes only 0.6 mW while requiring

100 ns for wakeup. Policies to move between these states will be important in achieving the low power potential of serial links.

Memory Network Power Management [95]

The HPCA 2017 paper by Jian et al. designs policies for exactly that purpose [95]. It explores a network of HMC-like memory devices and observes that link IO power, and idle link IO power are the main contributors to memory power, potentially contributing nearly 70%. This is not surprising for memory networks that are either minimally connected or where the network converges to a few links that connect to the processor. When the link to the processor is the bottleneck, many of the other network links will inevitably be underutilized, thus burning energy for always-on serial links. To reduce idle link IO power, Jian et al. assume various power-down modes—rapid on/off that consumes 1% power in off mode and requires 14 ns for wakeup, a DVFS mode that requires micro-seconds to modify voltages, and variable link width. Of these, the first is most effective. As a first step, Jian et al. show that each memory device can turn links off based on its own estimate of latency impact. The power savings can be doubled with more network-aware policies that share network state. Such policies ensure that busier routers remain powered on longer than less-busy routers; they also try to hide wakeup latencies for links used for responses; they also take upstream congestion into account when making power-down decisions.

Value Similarity

A number of recent papers have tried to reduce energy in memory interconnects by exploiting value locality/similarity. This is motivated by the fact that memory bus frequencies and interconnect power have been rising; for example, the bit rate doubled in going from GDDR5 to GDDR5X, but the energy per bit only reduced by 19%, thus causing a 63% increase in interconnect peak power [20]. Lee et al. [20] explain that the parallel interface in state-of-the-art GDDR5 and GDDR5X uses a circuit with *pseudo open drain* (POD). The current flow through the termination resistor leads to asymmetry—transmitting a 1 consumes 37% more energy than transmitting a 0. To reduce interconnect energy, it is important to not only reduce the number of bit toggles, but also transmit fewer 1 s. Lee et al. introduce a new way to encode a data transfer such that it transmits fewer 1 s, while having the side effect of lowering the toggle rate as well. This technique can be combined with *data bus inversion* (DBI)—every byte is sent in its original or flipped form so that there are more 0 s than 1 s. DBI, with an overhead of 1 wire per byte, is already used in baseline GDDR5 and GDDR5X. The technique of Lee et al. does not require any additional metadata or any change to the DRAM interface; it is a data encoding technique handled entirely by the memory controller.

The technique is based on the observation that words in a cache block are often similar to each other. The first word is therefore used as a base, and the second word is XOR-ed with the base before transmitting. The third word is XOR-ed with the unencoded second word before transmitting, and so on. Given the similarity in adjacent words, the encoded data block has abundant 0s. The authors conservatively use a large base word. This base word then undergoes additional encoding to increase 0s; it is recursively partitioned into two halves with the second half encoded by XOR-ing with the first half. Further, to handle the special case of zero strings, the authors introduce a mapping scheme where an unencoded zero word is transmitted as an encoded special word (with very few 1 s). To avoid aliasing, one other word has to swap its mapping with the zero word.

With the above encoding techniques in place, the authors reduce the number of 1 s by 35%. Combined with DBI, the authors reduce the number of 1 s by 48%, which eventually yields a 7% reduction in overall memory energy.

An earlier paper by Seol et al. [103] also observed value similarity in memory channel transfers. To find more opportunities to transmit 0s, they compare a word against (say) 64 recently transmitted words. A diff (XOR) with the most similar word is then transmitted. While the chances of finding a similar word are high, the implementation overheads are non-trivial— in particular, a table of recently transmitted words must be implemented in DRAM chips and the memory controller, and the memory channel must also transfer a pointer to the most similar word. CABLE is another recent approach that uses the processor's cache as a dictionary while implementing compression to reduce off-chip link bandwidth [104]. Related work includes "More is Less" [105], which converts a data word into a larger sparse code with fewer 1 s. While this lowers energy, it increases the bandwidth demand (just as a one-hot code reduces energy but takes more time). It must therefore be employed only when there is under-utilization in the bus.

Managing Bandwidth Across Memory Devices [106]

Future processors will likely be connected to a set of heterogeneous memory devices to form a deeper hierarchy. For example, a processor may be connected to a DRAM cache with an HBM interface, in addition to several high-capacity DIMMs with a DDR4 interface. The DRAM cache offers an uncontended latency that is only slightly smaller than the uncontended latency of the DDR4 interface. The gap is wider when bus contention is considered; the HBM interface can offer much higher bandwidth and fewer contention cycles. Gaur et al. [106] observe that the benefit of the DRAM cache is strongly influenced by the varying bandwidth pressures on the HBM and DDR4 interfaces. If the DRAM cache is very effective and yields a high hit rate, it also must deal with more contention cycles for the HBM interface, thus eroding the latency advantage of a DRAM cache hit. Gaur et al. therefore augment the DRAM cache with policies that can throttle the DRAM cache access rate so that bandwidth pressures on different memory channels are somewhat balanced. This is done with an epoch-based technique with counters to estimate memory channel contention cycles and the effectiveness of different throttling techniques. The throttling techniques include the following: avoiding filling the DRAM cache on a read miss, avoiding evictions from the previous level to the DRAM cache, avoiding the DRAM cache when the next level has a valid copy.

HMC Networks

A couple of papers by Kim et al. [107, 108] advocate for a network of CPUs/GPUs and HMC devices. They observe that this leads to efficient communication between CPUs, between GPUs, or between CPUs and GPUs, by avoiding intermediate routers and PCIe links. A network of HMCs is organized in the middle and surrounded by CPUs and GPUs. To increase path diversity and reduce network diameter, the authors recommend a sliced flattened butterfly topology.

In follow-up work, Zhan et al. [109] co-design the intra- and inter-HMC networks, using edge routers so non-local traffic can bypass the intra-HMC network; the intra-HMC network itself uses a dispersion topology, avoiding links between HMC vaults. The authors also use compression, but selectively employ it only for far-traveling packets to amortize the latency for compression/decompression. They also employ power gating for under-utilized links, while still keeping all memory devices connected and supporting deadlock freedom. Poremba et al. [110] improve memory network performance by adding express channels, an interposer-based package that emulates network concentration, heterogeneous DRAM/NVM networks, and distance-based arbitration. More recently, Ogleari et al. [111] propose a more scalable HMC network that reconfigures based on nodes that are powered off, and that adapts the path lengths, topology, and routing protocol.

New Interconnect Organizations for DIMMs

DIMMs are commodities that typically abide by DDR standards. In the past, DIMMs that have dared to innovate, e.g., FB-DIMM, have not had much success. Recent DDR standards have added innovations for energy and reliability, so it is conceivable that winds have shifted enough that DIMMs may evolve to incorporate popular ideas. For example, several works have highlighted the varied benefits of sub-ranks—splitting a conventional rank into finer-granularity sub-ranks can reduce energy, increase parallelism, and support efficient compression [17, 18, 67]. In the same vein, Yoon et al. [112] proposed a low-power DIMM organization (BOOM) where many low-frequency LPDDR chips and their interconnects are eventually multiplexed onto a standard DDR interface through an on-DIMM buffer chip. The Centaur memory buffer [113] is another example of a custom commercial DIMM that connects to the IBM Power 8 processor with serial links; it integrates a 16 MB cache and several memory management features typically handled by the processor. A recent paper [2] also introduces two new interconnect models to improve DIMM efficiency. First, it introduces an on-board buffer that enables cascaded DDR channels; this helps grow capacity without increasing the load on individual DDR channels. Second, it argues for narrow DDR channels and DIMMs as an alternative way to grow memory capacity without increasing the load on individual channel wires. The narrow DDR channel also enables lower frequency for on-DIMM interconnects and lower power, similar to the BOOM organization.

7.2.2 DISCUSSION

We've already talked about how a mix of serial and parallel interfaces may be an ideal trajectory for future memory systems in Section 7.1.3. Most enterprise and high-end platforms saturate available memory bandwidth, so these high-performance systems will likely use serial links where they can—the bandwidth advantage of serial links will usually offset their encoding latency disadvantage. Similarly, in large datacenters, some servers may have limited amounts of local memory, while a few memory blades in a rack may serve as a shared terabyte-scale repository of remote memory for applications that demand it [114]. Such terabyte-scale memory systems will no doubt use a variety of interconnects and require large amounts of data movement. More work is therefore required to define an optimal substrate, topology, and protocol. The underlying technology is still up in the air, especially with Micron recently discontinuing its HMC efforts. If 3D-stacked memory grows in popularity, the memory network could be largely constructed with serial links, which are clearly a superior technology. However, if we continue to use low-cost DRAM chips that need a parallel interface, hybrid serial-parallel networks will be required. No doubt, many of the ideas explored for on-chip networks [115] will also apply to these memory networks. As discussed above, some of the low-hanging fruit in these areas (adaptive throttling, data encoding) has already been picked. I anticipate that breakthroughs in this area will likely be triggered by new circuits and technologies—just as HMC-like devices injected new life in this area, we should keep an eye out for the next wave of exploitable technologies. As we'll discuss in Chapter 10, another promising area is the use of near-data computations embedded within the network to further reduce data movement.

CHAPTER 8

Memory Reliability

DRAM cells are implemented on tiny, narrow capacitors that lose their charge in a fraction of a second. Yet, DRAM cells are *extremely* reliable. However, put a trillion of these cells in a server, and hundreds of servers in a supercomputer, and the chances of a bit-flip are non-trivial. Therefore, memory is typically protected with redundant error correction codes (ECC) that take care of both error detection and recovery.

Today, DIMMs are available with and without ECC; both are commodities produced at high volume and low cost. DIMMs without ECC have a 64-bit interface, while ECC DIMMs have a 72-bit interface and 12.5% more DRAM chips. Historically, ECC DIMMs cost more than 12.5% more than non-ECC DIMMs. The memory controller takes care of storing/retrieving 72-bit code words and extracts the 64-bit data word before passing it to the processor.

The most commonly used ECC code is SECDED (Single Error Correct, Double Error Detect), which assigns an 8-bit code to every 64-bit data word. An ECC DIMM can support a variety of codes as long as the storage overhead is 12.5%. For example, extracting a 128-bit data word from a 144-bit code word can offer higher reliability than SECDED. Codes are usually constructed for every 64-bit data word transferred on the memory channel; they are usually not constructed for every 512-bit data burst. This is because a failed pin manifests as a single error in every 64-bit word, but as eight errors in every 512-bit burst; it is therefore easier to treat each 64-bit word (or more generally, each pin) independently.

Most large datacenters typically use ECC DIMMs with SECDED because manageability and availability are worth more than the incremental cost of ECC DIMMs. Some mission-critical installations demand higher levels of reliability; such systems support *"chipkill,"* i.e., they use codes that can recover even if an entire DRAM chip fails. "Double chipkill" refers to two DRAM chips failing. Chipkill correct memories typically use narrow DRAM chips, e.g., ×4 chips, so the code has to tolerate up to 4 consecutive failed bits.

In this chapter, we'll first provide some basics on memory error tolerance, followed by efforts to construct efficient codes for DRAM, and eventually NVMs.

8.1 BASICS OF DRAM ERRORS

DRAM Error Taxonomy

Memory errors are faults that impact observed program state; they are caused by a variety of factors: manufacturing defects, high-energy particle strikes, wearout, noise, etc. Failures are measured by the failure in time (FIT) rate, i.e., failures in every billion hours of operation. A hard error refers to a bit or pin that is permanently corrupted; a soft error refers to a bit flip caused by a transient event, i.e., a subsequent write and read involving that bit will operate correctly. It is worth noting that DRAM chips already include spare rows and columns; these are engaged in case of permanent faults detected during post-manufacture tests.

Memory errors are classified by their impact on the system and by their nature. In terms of system impact, the error can be a correctable error (CE, no harm done), detected but uncorrected error (DUE, quite damaging), and silent data corruption (SDC, catastrophic because the error persists and was never detected).

A number of studies [116–123] have quantified DRAM errors in production settings and have classified them based on their nature. The error types include: single-bit errors, double-bit (neighboring) errors, row failure, column failure, bank failure, pin failure, chip failure, multiple random bit errors. These empirical studies are extremely important because they not only guide us toward important problems, they also help quantify error rates for new approaches to detect/correct errors. We'll examine one recent study here to explain the research methodology and the conclusions they drive.

Empirical DRAM Error Study [117]

The study by Sridharan et al. [117] logged memory errors in two large supercomputers for over a year. One of the supercomputers was equipped with chipkill-detect (ability to detect complete failure in one DRAM chip), while the other was equipped with chipkill-correct memory (ability to correct complete failure in one DRAM chip). The DDR3 DIMMs used ×4 DRAM devices from three different vendors and the memory controller logged all encountered errors. By a wide margin, most errors are single-bit, split roughly evenly as soft and hard errors. There is a non-trivial number of multi-bit errors as well; most tend to have spatial correlation, e.g., an entire row being corrupted, potentially by row-hammer causes. The authors point out that parity checks for the address/command bus are vital. They also observed that for some of the vendors, the transient error rate increased with altitude (one of the installations was at elevation of nearly 8,000 feet), as much as 3× for one vendor. Because of the error logs with the strong error detection in these supercomputers, the authors were able to simulate the impact of weaker error detection codes like SECDED. They observed that a SECDED code would have an undetected error rate of 21.7 FIT per DRAM device, which translates to more than one undetected error daily for a supercomputer. Based on this, and projections to future memory capacities, the authors claim that SECDED is insufficient for modern large-scale installations and future systems may need codes even stronger than chipkill correct.

Scrubbing

Many systems will typically employ a background scrubbing operation that reads every memory block into the processor, and upon finding a correctable error, writes the corrected block back into memory. This is useful in identifying hard errors (which sometimes triggers

DIMM replacement in a datacenter), and in reducing the incidence of multiple (potentially uncorrectable) errors. This is a more expensive process than refresh because it involves data movement into the processor and ECC check; it is therefore performed at a slow rate of a few giga bytes per hour [116].

Single Bit Protection

Correction codes can be constructed in many ways. Most readers are probably familiar with RAID, which are simple error correction constructions because each disk controller takes care of error detection with cyclic redundancy checks (effectively a hash). Those solutions don't apply to DRAM because each DRAM chip does not have in-built error detection (that may be changing though). A DRAM error correction code must take care of both detection and correction. A 61-bit BCH code (Bose-Chaudhuri-Hocquenghem) can protect up to 1023 data bits from up to 6 errors [124]. But if such a code were used to protect 512 data bits, a pin failure would result in 8 errors, which would be uncorrectable. Therefore, the Hamming SECDED code [125, 126] is most widely used for DRAM. With the Hamming code, m code bits can protect k information bits, where $2^{m-1} \geq m + k$. Given this relationship, an 8-bit code can protect (single error correct, double error detect) data fields as large as 120 bits; a 7-bit code can protect data fields as large as 57 bits. Attaching an 8-bit code to every 64-bit data transfer is most compatible with the DDR standard.

Commercial Chipkill Implementations

The work of Kim et al. [127] provides an excellent background on codes and commercial implementations; we summarize some of that discussion here. Chipkill operates at the granularity of multi-bit symbols and uses Reed-Solomon (RS) codes. SSCDSD refers to single symbol correct, double symbol detect, i.e., it can handle all bits in a single symbol failing. Typically, all the bits produced by a DRAM chip on a clock edge constitute a symbol; if a chip fails, one symbol is corrupted. To correct an error in one symbol out of many, additional check symbols are required. Assuming a symbol size of 4 bits, 2 check symbols are enough for single symbol error detection and 3 check symbols are enough for single symbol error correction in up to 15 data symbols. Such a design is not a good fit for the DDR standard (only 60 data bits transmitted on a 72-bit channel in every beat), so 4 check symbols are typically used to create an SSCDSD code, capable of correcting an error in many more data symbols. Since the DDR standard with ECC demands a bus width in multiples of 72, the 4 check symbols are appended to 32 data symbols. This is therefore implemented with 36 ×4 chips spread across two 72-wide memory channels [128, 129]. This is a steep overhead in terms of limited rank/channel parallelism, overfetch to the processor, overfetch to the row buffers on several chips, and overall memory energy.

Newer systems [130, 131] use two beats of a ×4 chip to create an 8-bit symbol, i.e., a failed chip only impacts one symbol. Because of the larger symbol size (and intuitively, its greater information content), only 2 check symbols are enough to perform single symbol correction (SSC). This allows a DDR-compatible SSC implementation with $16 + 2 \times 4$ chips and a 72-bit channel. This code does not provide detection of two symbol failures, so other techniques [127, 130] are required to identify the common case scenario where at least some of the multi-symbol failures are previously identified errors. This also forms the basis for some double chipkill commercial approaches that downgrade to a weaker code after the first hard error has been detected [127].

DDR RAS Features

As technology scales, miniscule amounts of charge are being stored in DRAM cells that are very narrow/deep, and bits are subjected to noisy transmission over high-frequency links [132, 133]. The DRAM industry is acutely aware of the need for greater RAS (reliability, availability, serviceability) in future memories; this is reflected in a number of features that have found their way into the DDR4/DDR5 standards. DRAM chips are including support for in-DRAM error detection/correction [134]. Transmissions on the memory channel are supported by parity checks for the address/command bus and cyclic redundancy checks for the data bus [135]. The data bus also supports data bus inversion at byte granularity to reduce energy. And unlike DDR4 where ECC DIMMs are provisioned with 12.5% redundancy, DDR5 ECC DIMMs are provisioned with 25% redundancy. Such DDR5 DIMMs will support ranks with a width of 32 (data) + 8 (ECC) or 64 (data) + 16 (ECC). We will therefore see a number of future works that exploit this generous DDR5 metadata space for better RAS, compression, security, etc.

8.2   MEMORY RELIABILITY INNOVATIONS

There are a number of works that improve system metrics while providing high levels of memory reliability. We'll discuss some of the varied approaches here. The first three are examples of multitier codes; because of their greater complexity, they have not yet been embraced commercially. Most commercial implementations employ single-tier codes, i.e., the redundant code has enough information to both detect and correct a specified number of errors. This section also discusses a number of approaches to handle hard errors. While the early part of the decade focused on soft errors and better organized codes for DDR-based chipkill, recent interest has hovered on hard errors, in-DRAM ECC, and reliability for 3D-stacked memory.

Virtualized ECC, Yoon and Erez [136]

Yoon and Erez introduce the concept of tiered error protection for main memory. They break up the code into a first tier of error detection and a second tier of error correction. In the case of chipkill, two symbols are required for error detection (tier 1), while a third and possibly fourth symbol are used for error correction (tier 2). The tier 1 code can be accommodated on an ECC DIMM and is fetched along with the data block; shrinking the width of the code helps shrink the width of the rank and favorably impacts memory parallelism and energy. The tier 2 error correction code is stored separately in other pages managed by the operating system. In the uncommon case, when an error is flagged by tier 1, the tier 2 code is fetched. Of course, on every write, the tier 2 codes must be updated. To lessen this impact, they are cached in the LLC. By virtualizing error correction, the OS has the flexibility to implement stronger or weaker codes to match application requirements.

LOT-ECC, Udipi et al. [137]

Most chipkill correct techniques rely on RS codes and the requirement of 3 or 4 check codes. This places limitations on the width of each DRAM chip and the number of chips that must be ganged together for every data fetch. Udipi et al. introduce a new coding approach to provide nearly the same level of resilience while removing the above limitations. The technique uses three tiers of protection. The data block is scattered across 9 ×8 chips; each chip also has room for a hash that serves as a tier 1 error detection code. Once the erroneous chip is detected, its contents are corrected with RAID-like parity (tier 2 protection). The parity itself must be scattered across the DRAM chips and protected with a third tier of parity in case one of the DRAM chips has failed. The work therefore designs codes for memory system efficiency (parallelism and energy) instead of mapping unmodified codes to the memory system; the trade-off is higher storage overhead for codes and a higher SDC rate.

Multi-ECC, Jian et al. [138]

With Multi-ECC, Jian et al. fix some of the drawbacks of LOT-ECC while preserving its parallelism and energy benefits. The extra chip per rank is used for error correction, assuming that the error location has been detected. This error localization is performed with checksums for each DRAM chip. Given the potential storage overhead of strong checksums, these checksums are shared across multiple lines. These designs highlight the inter-play between various codes and the trade-offs in memory system metrics.

Bamboo ECC, Kim et al. [127]

Bamboo ECC builds on the chipkill strategy used in newer AMD chips and creates the state-of-the-art single-tier technique. Just as the AMD chipkill approach [130] uses 2 beats from a ×4 chip to create an 8-bit symbol, Bamboo ECC uses 8 beats from a pin to create an 8-bit symbol. The use of few pins and more beats makes this a tall/skinny "Bamboo" family of codes. The key advantage here is that it provides a more favorable and DDR compatible granularity. With just 2 redundant symbols (pins), an SSC code, equivalent to single-pin-correct (SPC), can be constructed for the 64 DDR data symbols (pins), i.e., SPC would not even require all the pins and memory provided by an ECC DIMM. The code strength can be increased: 4 redundant symbols can provide single-pin-correct-triple-pin-detect (SPC-TPD), 8 redundant symbols can provide quad-pin-correct (QPC), and 16 redundant symbols can provide octuple-pin-correct (OPC). With the 8 redundant pins on an ECC DIMM, one may start by implementing QPC, but as hard errors are discovered, graceful degradation to SPC-TPD and SPC can be implemented. The resulting system has very high reliability and favorable system performance/energy metrics. The only drawback with this flexible approach is that unlike prior chipkill techniques, the code has a large number of symbols (72), resulting in a 16× increase in transistors for decoding (which itself is a small fraction of the overall chip).

COP, Palframan et al. [70]

Palframan et al. show how error correction can be provided in the common case with non-ECC DIMMs, i.e., with no storage or bandwidth overhead. In doing so, they also augment our "architecture toolbox"—tricks that one may apply to other system problems. The authors attempt to compress each block; if the block is compressed, there is room to introduce Hamming ECC codes. The authors break up the 512-bit block into four regions, each using an 8-bit Hamming code to protect 120 bits of data. When a block is read and the four Hamming codes are correct, the block is assumed to be in compressed format and is accordingly decoded. If one Hamming code flags an error, the block is assumed to be compressed and the single error is corrected. If multiple Hamming codes are incorrect, the block is assumed to be uncompressed and the data is accepted as-is, i.e., it has no error protection and this is hopefully the uncommon case. Thus, ECC is being used to figure out if a block is compressed or not, avoiding separate metadata to track compressibility. If an uncompressed block coincidentally appears to have three or more correct Hamming codes, it is kept in cache and not written to memory—this prevents the block from being mistakenly interpreted as a compressed block. Finally, if uncompressed blocks need error protection as well, a separate memory region is reserved for their Hamming codes. A pointer at the start of the uncompressed block points to this Hamming code and the data bits displaced by the pointer.

Tolerating Hard Errors and In-Memory Error Correction

Nair et al. [133] observe that DRAM technology scaling results in deep cells that are more vulnerable to permanent faults. They argue for additional error protection that goes beyond ECC DIMMs and that can handle 100× higher bit error rates. The authors employ error profiling to create a Fault Map (stored in memory and cached on chip) that tracks if a 64-bit word has 0, 1, or 1+ errors. Words with errors are replicated in memory and accessed in case of errors that are uncorrected by ECC. Follow-up work from Son et al. [139] tried to address the same problem by adding an SRAM cache on each DRAM chip to service requests for faulty cells. On the other hand, FreeFault [140] and RelaxFault [141] lock faulty data in the LLC.

The impending unreliability of DRAM cells has prompted multiple proposals for in-DRAM error detection and correction [142, 143]. Udipi et al. [19] introduced in-DRAM checksums to flag an error; these were then augmented by RAID-like techniques across many DRAM chips. Nair et al. [144] evaluate a similar approach with in-DRAM ECC.

The work of Cha et al. [132] provides the best motivation for in-DRAM error checks. The authors show the effect of shrinking the cell capacitance for 4,500 DRAM dies manufactured with a Samsung 25 nm DRAM process. The results show a 100× increase in errors, high sensitivity to temperatures, and random distribution of errors. With circuit simulations, the authors also show an increase in random bit errors in future technologies because of variations in retention time and write recovery time. Samsung has also designed an LPDDR4 DRAM with support for single-bit error correction [134]. That prototype adds an 8-bit Hamming code to every 128-bit word, introducing an area, latency, and power overhead of 6%, 10%, and 3%, respectively. Cha et al. [132] also show that the 128 + 8 bit code is an ideal design point that is also compatible with rank-level ECC techniques. They note that on writes, the bits being modified on an individual chip are a subset of the 128 + 8 code; this requires a read-before-write operation that slows down writes. This has a small impact on performance. If a burst of writes is being performed and all 128 data bits on a chip are being modified, the read-before-write can be avoided.

Gong et al. [145] take an alternative route to reduce the overfetch and read-before-write overheads of in-DRAM ECC. The authors sidestep the in-DRAM correction and simply pass the redundant bits back to the memory controller. This is done by increasing the burst size from 8–10. With these additional bits at its disposal, the memory controller can implement stronger rank-level codes, thus improving reliability beyond what is possible by combining in-DRAM ECC and rank-level ECC.

Hard Errors in NVM

There is also a body of NVM work that deals with the high hard error rates caused by the limited endurance of NVM cells. Ipek et al. [146] dynamically replicated pages with hard errors. Schechter et al. [147] replace ECC with error correction pointers that store pointers to failed bits in a line and the correct values for those bits. Seong et al. [148] note that in some NVMs, the hard errors are stuck-at faults; the cell is therefore still usable as long as the word is selectively written in its raw form or in its inverted form. FREE-p [149] uses an erroneous line to store (replicated) pointers to an error-free version of the line. NVMs are also vulnerable to correlated soft errors, e.g., because of resistance drift phenomena; Awasthi et al. [150] design scrub policies to efficiently detect impending uncorrectable states. More recently, Zhang et al. [151] design two tiers of codes to deal with NVM errors; traditional DRAM-like codes are used at runtime, but a second set of codes at larger data block granularity are used when the error rate is high or when recovering after a reboot. Treating long-term errors and runtime errors differently in this way leads to overall lower storage overhead.

Heterogeneous Memory Trade-Offs

Gupta et al. [152] make the observation that tiered memory systems with in-package 3D-stacked memory and off-package DDR memory can introduce performance/reliability trade-offs. The premise is that in-package memory can have higher failure rates because of TSVs and it suffers from higher bandwidth/storage overheads from error correction codes [153]. It is therefore viewed as the higher performing, but less reliable tier of memory. The authors make the case that the ideal data placement policy is one where the in-package memory contains hot pages with low-risk data. They show that nearly 40% of pages are hot and have low architectural vulnerability factor (AVF, a measure of risk to the program). Gupta et al. design heuristics to steer such pages to the in-package memory, yielding an increase in reliability while suffering minor performance losses.

8.3 DISCUSSION

DRAM chip error rates are expected to increase as technology continues to shrink [117, 133]; soft and hard errors are expected to be high for some non-volatile memories as well. Memory reliability will therefore be vital in the future [117]. Reliable memory systems deal with various trade-offs: storage overhead, bandwidth overhead, granularity/parallelism for data access, error rates, etc. Most recent innovations in this area are therefore trying to provide high levels of error tolerance with minimal overheads. Single-tier reliability also appears more viable for industry. Students looking for impactful problems may want to focus more on the trade-offs involved in chipkill or double-chipkill systems. While the overhead of chipkill reliability has shrunk significantly for commodity DDR, it continues to be a challenge for memory systems with fine-grained data access, including emerging 3D-stacked memories. Similarly, the overhead of updating in-DRAM ECC remains an open problem.

CHAPTER 9

Memory Refresh

Most computer scientists have no idea that the entire contents of their memory are refreshed more than a dozen times every second. The DRAM retention time is very small—64 ms. So every one of the billions of DRAM cells in any system must be read and restored every 64 ms. At first, this appears to be a very burdensome process. Thanks to some super engineering, in most technologies, the overheads of refresh are fairly small. But the refresh overhead is projected to grow as cells continue to shrink. In anticipation of this high overhead refresh, a number of alleviating techniques have been proposed, which we will discuss in this chapter.

9.1 REFRESH BASICS

Frequency of Refresh Operations

The charge on a DRAM cell weakens over time. The DDR standard requires every cell to be refreshed within a 64 ms interval, referred to as the *retention time*. Typically, most cells can retain their charge for much longer than that. But with a conservative DDR specification, DRAM vendors can better tolerate the inevitable weak cells on a chip. A more aggressive retention time (say, 128 ms) would have lowered the refresh burden, but would have resulted in more failed cells per chip and lower yield. At temperatures higher than 85°C (referred to as extended temperature range), the retention time is halved to 32 ms to accommodate a higher leakage rate per cell.

The refresh of a memory rank is partitioned into 8,192 smaller refresh operations. One such refresh operation has to be issued every 7.8 μs, which is simply 64 ms/8192. This 7.8 μs interval is referred to as the *refresh interval, tREFI*. The DDR3 standard requires that eight refresh operations be issued within a time window that equals $8 \times tREFI$. This gives the memory controller some flexibility when scheduling refresh, i.e., some of the refresh operations can be delayed if the memory controller is busy performing other more urgent operations. Refresh operations are issued at the granularity of an entire rank in both DDR3 and DDR4. Before issuing a refresh operation, the memory controller precharges all banks in the rank. It then issues a single auto-refresh command to the rank. DRAM chips maintain a row counter to keep track of the last row that was refreshed—this row counter is used to determine the rows that must be refreshed next. Thus, the low-level details of refresh are managed internally by the DRAM chips, while the memory controller is only responsible for periodically reminding the chips to perform their next refresh. The DRAM chip can also enter self-refresh mode where many of the clock/IO circuits are disabled for low power, and an internal analog timer is used to perform refreshes without memory controller intervention.

tRFC and Recovery Time

Upon receiving a refresh command, the DRAM chips enter a refresh mode that has been carefully designed to perform the maximum amount of cell refresh in as little time as possible. During this time, the current carrying capabilities of the power delivery network and the charge pumps are stretched to the limit. Hence, no other read/write operations can be performed on the chip while the refresh is underway. The operation lasts for a time referred to as the *refresh cycle time, tRFC*. Toward the end of this period, the refresh process starts to wind down and some recovery time is provisioned so that the banks can be precharged and charge is restored to the charge pumps. Providing this recovery time at the end allows the memory controller to resume normal operation at the end of tRFC. Without this recovery time, the memory controller would require a new set of timing constraints that allow it to gradually ramp up its operations in parallel with charge pump restoration. Since such complexity can't be expected of every memory controller, the DDR standards include the recovery time in the tRFC specification. As soon as the tRFC time elapses, the memory controller can resume its otherwise most current-intensive operations—issuing four consecutive Activate commands to different banks in the rank.

Refresh Penalty

On average, in every tREFI window, the rank is unavailable for a time equal to tRFC. So for a memory-bound application on a 1-rank memory system, the percentage of execution time that can be attributed to refresh (the refresh penalty) is tRFC/tREFI. In reality, the refresh penalty can be a little higher because directly prior to the refresh operation, the memory controller wastes some time precharging all the banks. Also, right after the refresh operation, since all rows are closed, the memory controller has to issue a few Activates to re-populate the row buffers. These added delays can grow the refresh penalty from (say) 8% in a 32 Gb chip to 9%. The refresh penalty can also be lower than the tRFC/tREFI ratio if the processors can continue to execute independent instructions in their reorder buffers while the memory system is unavailable. In a multi-rank memory system, the refresh penalty depends on whether ranks are refreshed together or in a staggered manner. If ranks are refreshed together, the refresh penalty, as above, is in the neighborhood of tRFC/tREFI. If ranks are refreshed in a staggered manner, the refresh penalty can be greater. For example, in a 4 rank system, portions of memory are unavailable for a fraction of time that equals 4×tRFC/tREFI. Since all threads typically scatter their requests across all ranks in the system, all threads are vulnerable to stalls during these periods of unavailability. In spite of the potentially lower performance with staggered refresh, it is is frequently employed because it reduces the memory's peak power requirement. Note that refresh is the most current-intensive operation in DRAM and we ideally don't want all ranks to be refreshing at the same time.

Misconceptions Regarding Refresh

Now consider how a few rows in all banks are refreshed during the tRFC period. As DRAM chip capacities increase, the number of rows on the chip also increases. Since the retention time (64 ms) and refresh interval (7.8 μs) have remained constant over the years, the number of rows that must be refreshed in every refresh interval has increased. In modern 4 Gb chips, eight rows must be refreshed in every bank in a single tRFC window. Some prior works have assumed that a row refresh is equivalent to an Activate+Precharge sequence for that row. Therefore, the refresh process was assumed to be equivalent to eight sequential Activate+Precharge commands per bank, with multiple banks performing these operations in parallel. However, DRAM chip specifications reveal that the above model over-simplifies the refresh process. First, eight sequential Activate+Precharge sequences will require time = 8 × tRC. For a 4 Gb DRAM chip, this equates to 390 ns. But tRFC is only 260 ns, i.e., there is no time to issue eight sequential Activate+Precharge sequences and allow recovery time at the end. Also, parallel Activates in eight banks would draw far more current than is allowed by the tFAW constraint. Second, the DRAM specifications provide the average current drawn during an Activate/Precharge (IDD0) and Refresh (IDD5). If Refresh was performed with 64 Activate+Precharge sequences (64 = 8 banks × 8 rows per bank), we would require much more current than that afforded by IDD5. Hence, the refresh process uses a method that has higher efficiency in terms of time and current than a sequence of Activate and Precharge commands.

The Details of a Refresh Operation

While the specific details of a refresh operation may vary across chips and vendors, below are some insights that help explain some of the secret sauce behind the latency and current efficiency of refresh. A bank is designed with a number of subarrays. For example, a bank may have four subarrays, of which only one is accessed during a regular Activate operation. This observation also formed the basis for the subarray-level parallelism (SALP [82]) idea of Kim et al. During a refresh operation, the same row in all four subarrays undergo an Activation and Precharge. In this example, four rows worth of data are thus being refreshed in parallel within a single bank. The current requirement for this operation is not 4× the current for a regular Activate; by sharing many of the circuits within the bank, the current does not increase linearly with the extent of subarray-level parallelism. Thus, a single bank uses the maximum allowed current draw to perform parallel refresh in a row in every subarray; each bank is handled sequentially (refreshes in two banks may overlap slightly based on current profiles), and there is a recovery time at the end. In short, a refresh operation is a finely-tuned and efficient process, one that is best left in the hands of DRAM circuit designers.

Refresh in DDR4

An important change introduced in DDR4 devices is a Fine Granularity Refresh (FGR) operation [8]. FGR-1x can be viewed as a regular refresh operation, similar to that in DDR3. FGR-2x partitions the regular refresh operation into two smaller "half-refresh" operations. In essence, the tREFI is halved (half-refresh operations must be issued twice as often), and the tRFC also reduces (since each half-refresh operation does half the work). FGR-4x partitions each regular refresh operation into four smaller "quarter-refresh" operations. Since each FGR operation renders a rank unavailable for a short time, it has the potential to reduce overall queuing delays for processor reads. But it does introduce one significant overhead. A single FGR-2x operation has to refresh half the cells refreshed in an FGR-1x operation, thus potentially requiring half the time. But an FGR-2x operation and an FGR-1x operation must both incur the same recovery cost at the end to handle depleted charge pumps. DDR4 projections for 32 Gb chips show that tRFC for FGR-1x is 640 ns, but tRFC for FGR-2x is 480 ns. The overheads of the recovery time are so significant that two FGR-2x operations take 50% longer than a single FGR-1x operation. Similarly, going to FGR-4x mode results in a tRFC of 350 ns. Therefore, four FGR-4x refresh operations would keep the rank unavailable for 1400 ns, while a single FGR-1x refresh operation would refresh the same number of cells, but keep the rank unavailable for only 640 ns. The high refresh recovery overheads in FGR-2x and FGR-4x limit their effectiveness in reducing queuing delays. In fact, some results [154] show that FGR is never better than traditional refresh. Given this more recent data about the high cost of refresh recovery overheads, techniques like Refresh Pausing [155] are also likely to be ineffective.

Refresh in LPDDR2

LPDDR2 also provides a form of fine granularity refresh. It allows a single bank to be refreshed at a time with a REFpb command (per-bank refresh). For an eight-bank LPDDR2 chip, eight per-bank refreshes handle as many cells as a single regular all-bank refresh (REFab command). A single REFpb command takes much more than $1/8^{th}$ the time taken by a REFab command—REFab takes 210 ns in an 8 Gb chip and REFpb takes 90 ns. Similar to DDR4's FGR, we see that breaking a refresh operation into smaller units imposes a significant overhead. However, LPDDR2 adds one key feature. While a REFpb command is being performed in one bank, regular DRAM operations can be serviced by other banks. DDR3 and DDR4 do not allow refresh to be overlapped with other operations. Micron datasheets indicate that a REFpb probably has a current profile that is somewhere between a single-bank Activate and a REFab.

Refresh Scaling Trends

Most papers on DRAM refresh have been motivated by the following projections. Refresh overheads have been expected to increase dramatically as DRAM chip capacities increase. In recent years, DRAM capacities have increased thanks to stacking, and not because of scaling, so the projected higher tRFC values haven't emerged yet. Table 9.1 shows the expected scaling trend [8], including the tRFC/tREFI ratio that is indicative of the refresh performance overhead. Note that the refresh energy overhead is typically higher than the performance overhead since the current draw during refresh is higher than the average current draw during regular operation. The table also shows the tRFC and tREFI for the various FGR modes.

Table 9.1: Refresh latencies for high DRAM chip capacities [8]

Chip Capacity (Gb)	t_{RFC}/t_{REFI}	t_{RFC} (ns)	$t_{RFC_{1x}}$ (ns)	$t_{RFC_{2x}}$ (ns)	$t_{RFC_{4x}}$ (ns)
8	4.5%	350	350	240	160
16	6.2%	480	480	350	240
32	8.2%	640	640	480	350
t_{REFI}		7,800	7,800	3,900	1,950

The number of cells that must be refreshed in every tRFC increases linearly with capacity. Therefore, we also see a roughly linear increase in tRFC. In future 32 Gb chips, the tRFC is expected to be as high as 640 ns, giving a tRFC/tREFI ratio of 8.2%. At high temperatures, this ratio doubles to 16.4%. The performance overhead can be even higher when employing staggered refresh across ranks. tREFI will also reduce if DRAM cell capacitances reduce in the future [133]. In a 3D stacked package, the number of cells increases without a corresponding increase in pin count and power delivery [85]—this too can result in a high tRFC.

9.2   REFRESH INNOVATIONS

9.2.1   EMPIRICAL STUDIES AND RETENTION TIMES

In the past decade, a number of empirical studies have been performed to understand DRAM cell retention times [156–158] and DRAM cell parameter/latency variations [90, 91, 93]. An early characterization [156] showed that very few DRAM cells on a chip are likely to lose data if the standard refresh interval of 64 ms is quadrupled. In short, there are a few weak cells that impose a very conservative refresh frequency on the billions of DRAM cells on a chip.

Assuming that these weak cells can be identified, mechanisms can be instituted to conservatively handle weak cells, while relaxing the constraints on strong cells. RAIDR [159] does exactly this. It assumes that rows containing weak cells can be identified; rows are then grouped into two bins depending on whether they contain weak cells or not. This grouping can be efficiently performed with Bloom filters. The two groups then employ different refresh intervals, thus significantly lowering refresh overheads. In an early work, Venkatesan et al. [160] also propose that the OS can preferentially allocate pages with long retention times.

However, in their follow-up work, Liu et al. [157] observe that identifying weak cells is not as easy as expected. With empirical measurements on hundreds of DRAM chips using an FPGA infrastructure, they show that cell retention time does not remain constant for a cell. In particular, they identify two new phenomena—*variable retention time* (VRT) that causes a cell's retention time to evolve over time, and *data pattern dependence* (DPD) where a cell's retention time is a function of data stored in nearby cells. Both phenomena make it harder to bin a cell as weak or strong. With sufficient profiling effort, and by sampling with a number of data patterns, the effects of DPD may be captured. But the existence of VRT implies that such profiling must be performed continuously, i.e., the profiling overheads may be non-trivial.

In more recent work, Patel et al. [158] try to lower the profiling overheads. Their goal is to capture a vast majority of the weak cells (over 99% coverage) with as little effort as possible. With an empirical characterization of DRAM chips, they show that this is possible by checking for retention time failures at a much higher refresh interval. This approach is so effective at finding the weak cells that it achieves sufficient coverage with fewer iterations than the baseline approach that uses a shorter refresh interval. However, the longer refresh interval may result in each iteration taking longer. To further complicate the profiling, some non-weak cells in the above approach get marked as weak, i.e., it results in more false positives. Thus, designing an effective profiling mechanism requires navigation of a complex trade-off space. Even with such continuous profiling, because of VRT, a cell categorized as strong may evolve and become weak; to overcome any resulting errors from these false negatives, we may have to rely on strong ECC measures. The efficient characterization and handling of DRAM cells therefore remains an open and complex problem.

In addition to addressing the profiling overheads, a few efforts have tried to directly alleviate the effects of DPD and VRT. Note that these proposals are in the context of a system where some rows are refreshed at slower rates than others. Khan et al. [161] try to detect vulnerability to DPD-caused errors during writes and lower the overheads of this detection by only targeting writes to specific pages. Qureshi et al. [162] increase the refresh rate for a row if a VRT error shows up in that row. Orthogonally, stronger ECC codes can be allocated to error-prone regions of memory [163, 164].

Bhati et al. [165] note that techniques that elide refresh for certain memory regions need additional support within the DRAM chip. The DRAM chip has its own book-keeping counters and automatically refreshes the next set of rows on every refresh command; the DRAM chip must therefore recognize a dummy refresh command that advances the internal counters without performing refresh. This allows the memory controller to retain its sophisticated decision-making and efficiently dictate variable refresh rates in each memory region.

In addition to the profiling studies of commercially available DRAM chips, we have previously discussed the work of Cha et al. [132]. That work manufactures DRAM dies with lower cell capacitance and shows its high impact on error rates and its high sensitivity to temperature. With additional simulation based evaluations, that study concludes that the discussed problems will only worsen as technology continues to scale.

9.2.2   ALTERNATIVE HARDWARE TECHNIQUES

We've seen above that a whole body of refresh research tries to apply different refresh rates to different rows in memory, but requires a complex supporting ecosystem. We will now discuss some alternative techniques that also reduce or hide refresh overheads.

Some of the early work in this area includes the following. Elastic Refresh schedules refresh during predicted idle times [166]. If a rank has been idle for a while, a refresh is scheduled to that rank; after eight refresh intervals, any pending refreshes are forcibly performed. Smart Refresh avoids refresh of rows that have been recently read/written [167]. It requires additional book-keeping within the memory controller and changes to the DRAM refresh interface.

More recently, a couple of papers [168, 169] overlapped regular DRAM operations with a lightweight refresh operation. As discussed in Section 9.1, LPDDR2 introduces one such lightweight refresh operation—the per-bank refresh. Unlike DDR's FGR modes, the per-bank refresh allows other banks to concurrently service regular DRAM operations. It can thus serve as a light (and slow) background refresh while the DRAM operates with slightly lower parallelism—this idea clearly has merit and with more innovations, it may serve as an elegant and commercially viable approach to hide refresh overheads. Chang et al. [169] project that a similar per-bank refresh in DDR would perform about 5% better than the conventional all-bank refresh. To boost this improvement, Chang et al. [169] and Zhang et al. [168] introduce better scheduling of refresh and regular operations; for example, Chang et al. try to schedule per-bank refreshes during a write drain. Further, they allow some subarrays in a bank to work on refresh while other subarrays can service regular operations. As explained earlier, refresh is a carefully engineered process and some of the mechanisms in these papers may not be compatible with a baseline refresh process. For example, if the baseline refresh process concurrently involves all subarrays in a bank to boost efficiency, it may not be possible to easily implement additional parallelism within a bank.

The recent Nonblocking Refresh technique [170] is a clever implementation of a lightweight background refresh while allowing concurrent regular operations. It is an interesting application of erasure codes to hide the latency of accessing unavailable data. In chipkill correct memory systems (or more generally, in systems where redundant bits are stored to enable error detection/correction), 18 DRAM chips may be involved in fetching one data block. With erasure codes, 16 of these chips contribute the actual data, one chip has information for error detection, and one chip has information for error correction once an entirely faulty chip has been identified. When a refresh command is issued to such an 18-chip rank, in round-robin fashion, one of the chips proceeds with refresh, while the other 17 chips skip the refresh (this requires simple logic within the DRAM chip). This rank is not unavailable at this time; it can service regular operations; but one chip that is busy with refresh, fails to provide data. With erasure codes, error detection has high latency, while error correction only requires a few cycles. Since a known chip has failed to provide data, the correction algorithm is used to quickly reconstruct the missing data. With all information now known, the error detection algorithm can be applied to provide the same fault tolerance guarantees as before. When errors are detected, the correction is slowly performed after the refresh has completed.

With this Nonblocking Refresh approach, refresh has a minimal impact on read latency and bandwidth. It does require more frequent refresh commands per rank, but all chips in the rank ignore a large fraction of these refresh commands. Writes are more problematic because they do require an update of every chip in the rank. To handle writes, only some of the ranks are involved in Nonblocking Refresh at a time, while other ranks are not involved in any refresh; the memory controller then favors writes to these other ranks during the write drain process.

We have thus seen two examples of a "lightweight refresh"—one that keeps only a few banks busy in a rank and one that keeps only a few DRAM chips busy in a rank.

9.2.3   ALTERNATIVE SOFTWARE TECHNIQUES

System software can also play a role in lowering refresh overheads. Isen et al. [171] observe that pages that have been freed need not be refreshed. In a similar vein, RAPID [160] steers the OS toward using pages with long retention times. Flikker [172] tries to identify application pages that can tolerate lower fidelity. Such pages are then refreshed at a lower frequency. This can be an effective technique in the toolkit for approximate computing.

More recently, Shevgoor et al. [154] show that OS page mapping is more effective at hiding refresh overheads than many of the hardware-based techniques. Most DRAM refresh papers have ignored the issue of simultaneous vs. staggered refresh. From a system vendor perspective, staggered refresh across ranks is desireable because it lowers the memory peak power requirement [8]. But with staggered refresh, for a large fraction of execution time, a subset of memory is unavailable. Since applications typically spread their pages across all ranks, all applications are potentially affected when a portion of memory is unavailable. As a result, staggered refresh ends up amplifying penalties from refresh. Shevgoor et al. propose that applications should cluster their working sets to a few ranks instead of spreading them out. This makes them less vulnerable to significant refresh penalties, while suffering from slightly lower memory level parallelism.

A similar OS approach was also used by Kotra et al. [173] to localize an application to a few banks. They then schedule LPDDR-style per-bank refreshes and processes in a symbiotic manner. The memory controller avoids refreshing banks that are currently being used by active processes and likewise, the OS scheduler picks processes to avoid banks that need the next refresh. The per-bank refreshes can thus proceed without slowing down currently scheduled applications that have all their data in other banks.

9.2.4   LEVERAGING CHARGE IN CELLS

Refresh influences the charge in a DRAM cell; the amount of charge then influences the latency of various DRAM operations. To keep the memory controller simple, uniform latencies have been assumed for each DRAM operation. But a few works have pointed out that performance can be improved if DRAM operations were charge aware and/or refresh aware.

This was first noted by Shin et al. [174]. They introduce a non-uniform access time (NUAT) memory controller that organizes the DRAM rows into 4 bins based on how recently they have been refreshed. The rows that have been refreshed most recently exhibit the best latency for an Activate operation—given the high charge in the cells, it takes less time for the sense amplifiers to detect the charge. The memory controller thus imposes one of four different timing parameter values (for tRCD, tRAS, tRC) based on the bin being accessed. The resulting performance improvement is small, but a key feature here is that DRAM chips are unmodified and the memory controller changes are minimal. Shin et al. further modify the scheduler to use the timing parameters per bin and performance counters to decide when to switch between open-page and close-page policies. The basic observation of NUAT was later extended by Hassan et al. [175] to also lower the timing parameters for rows that have been recently accessed. The ChargeCache tracks these recently accessed rows at the memory controller.

While NUAT exploits a latency opportunity at the start of a DRAM operation (the Activate), a subsequent proposal, Restore-Truncation [176], exploits the latency opportunity at the tail end of a DRAM operation. After a row has been read by the sense-amps, the charge on the cells has to be restored before the bank can be precharged. In conventional DRAM, a full restore is performed. Zhang et al. [176] observe that instead of a full restore, a partial restore may be enough to keep the cells "alive" until their next refresh. In other words, a row must be restored to the partial charge state that it would have been in, had it never been activated. In fact, the nature of the charge decay curve allows saving a significant number of cycles with this approach. Restore truncation also helps reduce energy. The implementation is simple (localized to the memory controller) and very similar to that of NUAT. Rows are organized into four bins based on their expected next refresh; each bin uses different timing parameter values to truncate its restore operation early. Zhang et al. also observe that there is a trade-off between refresh and restore truncation. In DRAM systems with variable refresh rates, increasing the refresh frequency increases the refresh overhead, but provides more opportunities for restore truncation. To exploit this trade-off, they briefly increase the refresh rate for portions of DRAM depending on the access frequency to that region. Follow-up work by Wang et al. [177] observes that restore truncation can also be useful when the row is likely to be activated in the near future. The authors design a predictor for upcoming activations with an accuracy of 98%. They identify a level of restoration such that the subsequent activate can also enjoy charge-aware lower latency.

9.3 DISCUSSION

While projections for refresh in future technologies appear dire, the past few years have also seen significant breakthroughs in managing refresh. Recent techniques—non-blocking refresh, per-bank refresh, OS mapping/scheduling—have proved to be very effective at hiding the latency impact of refresh. Per-bank refresh is expected to be a part of DDR5. Non-blocking refresh, if implemented, can be especially helpful because it eliminates an important source of latency uncertainty for datacenter workloads and systems. It may therefore be difficult to take big performance strides with smarter refresh policies. It is noteworthy that hiding the energy overhead of refresh is much more difficult—a fixed amount of refresh work has to be completed within a deadline, regardless of how that work is scheduled. The primary approach to reduce refresh energy is to elide refreshes to rows under certain conditions. Such approaches may be unattractive to industry because of book-keeping complexities. The techniques that exploit variable charge in cells may also be challenging to implement commercially given their complexity and modest performance improvements.

CHAPTER 10

Near Data Processing

Over the decades, researchers have seen the value in unifying compute and memory, but market forces have typically conspired to separate the two. In the 1990s, *Processing-in-Memory* (PIM) was in vogue. The grand plan was to design DRAM chips that not only had arrays of DRAM bits, but also some compute units so you didn't have to ship data all the way to the processor for simple operations. Interest in this area fizzled out once it became evident that DRAM manufacturers weren't interested. There were just too many challenges in implementing and selling PIM chips in a DRAM market driven by low cost and small margins.

But the last few years have seen a revival of interest in this topic. Using 3D stacking and NVM technologies as catalysts, many researchers have again made the case for unifying compute and memory. Within five years of this revival, industry prototypes have started to emerge, e.g., from IBM [178, 179], EMU Technologies [180], HPE [181], and Micron [182]. These span a range of implementations. The system from EMU [180] uses a large number of cores on many boards and uses a programming model where threads migrate to the location of data. ConTutto is an IBM platform that allows an FPGA-based DIMM to plug into a memory channel, enabling emulated experiments and in-line data processing [179]; another prototype from IBM and the University of Illinois uses an FPGA on a DIMM to emulate an ethernet-connected independent system [178]. At the other end of the spectrum, compute has been moved all the way into memory arrays: Micron's Automata processor adds specific operations in memory arrays [182] and prototypes of resistive memories have demonstrated analog dot-product operations within the arrays [181].

Lessons have been learned from past failures. There appears to be a stronger focus on killer apps and programming frameworks. Initial steps were careful to distance themselves from the work of the 1990s. The field was re-branded with an emphasis on *near-data* instead of *in-memory*. *Near Data Processing* (NDP) essentially implies that the compute and memory are placed on different dies; such specialization typically yields an efficient compute die and an efficient memory die. The dies can be connected in a variety of ways, including with TSV-based 3D stacking, or with an interposer (what is considered 2.5D stacking), or with direct links on a board.

As the weariness around PIM has worn off, groups have also started dipping their toes back into new incarnations of PIM that tightly integrate compute and memory on the same die. Such *in-situ processing* has been enabled by new materials and circuits for memory cells. We'll next cover both forms of unified compute and memory: NDP and in-situ.

10.1 NDP IMPLEMENTATIONS

We'll examine two forms of NDP, one where 3D stacking is leveraged to maximize bandwidth at relatively high cost (Section 10.1.1), and the other that avoids 3D stacking and simply relies on a new DIMM architecture (Section 10.1.2).

10.1.1 3D STACKED ARCHITECTURES

The Micron HMC [1] integrated DRAM dies and a logic die in a single 3D-stacked package. The logic die primarily has memory controller and routing functionality. The HMC was the first "disruptive" memory product in years; it favored performance over cost, it offered unprecedented levels of bandwidth and parallelism, and it was one of the first products to adopt 3D stacking. It was a demonstration of how logic and memory could be combined in a single package with high internal bandwidth. It also signaled that the memory industry was willing to enter the NDP arena, which it had abandoned a decade earlier.

The introduction of the HMC therefore sparked a number of projects on NDP. Even though the HMC didn't have a full-fledged processor on it, the obvious question was: what is the potential benefit from adding one or more processors to the HMC, and what would those processors look like?

NDC, Pugsley et al. [183]

An analysis by Pugsley et al. [183] shows that the key benefit of adding processors in an HMC is not the lower latency for memory access, but the high bandwidth offered by the TSVs in the HMC. The second major benefit is that as HMCs are added to the system, the processing power and the available memory bandwidth scale up linearly. This leads to two observations. First, NDP primarily benefits applications with high levels of parallelism. As a potential killer app, Pugsley et al. identify Spark workloads, i.e., the programmer-friendly and highly parallel MapReduce framework where a significant portion of the dataset is placed in DRAM. Second, if the goal is to maximize throughput at a given power budget, the processing cores in the HMC must be optimized for low energy per instruction. Therefore, the HMC's logic die should accommodate as many simple in-order cores as is allowed by the area and thermal budgets. The combination of the above features yields a 15× speedup with the NDC (Near Data Computing) architecture, relative to an iso-core baseline that has high-bandwidth HMCs but no in-HMC processing.

TOP-PIM, Zhang et al. [184]

TOP-PIM considers a similar architecture as NDC, but explores a larger design space of both processing cores and workloads. Key novelties include the use of GPU-style cores, a reliance on existing GPU software frameworks, a machine learning-based simulation methodology, and evaluations of graph and HPC workloads. They report a nearly 7× reduction in energy, but relatively small performance improvements, compared to GPUs.

AMC, Nair et al. [185]

The Active Memory Cube (AMC) architecture from IBM confirmed some of the observations made by the NDC and TOP-PIM architectures. The key novelties in the AMC design were its focus on exascale workloads/requirements, a vector architecture for the near data processor, avoiding unnecessary units like caches, a scatter-gather unit, predication, and compiler scheduling across the many vector lanes to further reduce energy per operation. Nair et al. [185] identify exascale workloads as killer apps because many (MPI) applications tend to stream through data with little immediate reuse. On workloads like matrix-multiply, they show nearly an order of magnitude improvement over existing IBM Blue Gene/Q installations in terms of throughput/watt, in large part because of higher in-AMC bandwidth and the higher parallelism afforded by vector lanes in multiple AMCs.

NDA, Farmahini-Farahani et al. [186]

The NDA design explores different approaches to 3D-stack an accelerator chip on a standard DRAM chip. The authors explore the impact of three different interfaces, each moving closer to the arrays to offer higher bandwidth to the accelerator; even the most invasive design imposes an area overhead of under 8%. The authors envision several such 3D-stacked packages on a DIMM that is accessed by the processor with a DDR protocol; the processor uses registers on the DRAM chip to hand access control to the 3D-stacked accelerator when required. CGRAs [187] are used to implement the accelerators and target a number of parallel applications, yielding an overall speedup of 1.67×.

Analytics, Gao et al. [188]

Gao et al. [188] extend the above works by adding support for communication among NDP cores with a combination of hardware and software. By allowing NDP cores to remotely access pages mapped to other vaults, the authors show improvements even when coarse-grained parallelism is limited. Per-core few-entry TLBs for large pages and VM protections are leveraged so that the OS/runtime can implement page granularity coherence among the many cores. The authors also introduce a pull-based prefetching model to lower the inter-core communication cost. These communication primitives yield a 2.5× speedup over the NDC baseline for MapReduce, graph, and deep neural network workloads. Gao et al. also show that additional software optimizations can further exploit NDP, e.g., edge-centric versions can better exploit spatial locality than vertex-centric versions of graph workloads.

10.1.2 TIGHT COUPLING ON A DIMM

While the approaches discussed above take advantage of 3D-stacked devices, there is no guarantee today that such devices will be cost-effective and commercially popular. In fact, Micron announced in 2018 that it will be moving away from HMC architectures. It is therefore worth pursuing an alternative line of research that tries to achieve NDP benefits without investing in 3D stacking, but by using commodity processor chips and commodity DRAM chips. The anticipated product from EMU Technology also falls in this class.

NDC-Module, Pugsley et al. [189]

As a follow-up to the HMC based NDC model described earlier, Pugsley et al. [189] introduced NDC-Module where a DIMM is populated with low-energy processor chips, each directly connected to its dedicated LPDDR2 chip with a 32-bit parallel interface. Unlike the baseline where a block is interleaved across an entire rank, a block must now be mapped to a single DRAM chip. Many such DIMMs together form a distributed framework that provides significant speedups for MapReduce workloads. However, because of the lower overall memory bandwidth on these DIMMs, the NDC-Module design only achieves about half the speedup possible with an HMC based NDC architecture.

Chameleon, Asghari-Moghaddam et al. [190]

The Chameleon architecture provides a more detailed exploration of the design space, while also dealing with commercial realities: how can NDP be implemented in the context of DDR4 LRDIMMs, how can DIMMs be used in both NDP and non-NDP mode, etc. It first notes that DDR4 LRDIMMs (Section 3.1) have a specific organization that reduces interconnect length/load, thus catering to the signaling requirements of various interconnects. A central chip on the LRDIMM receives commands and addresses from the DDR4 bus and then communicates these to all DRAM chips on the LRDIMM. A data buffer chip on the LRDIMM is used to exchange data between the DDR4 bus and a DRAM chip in each rank; multiple such data buffer chips are required and are scattered across the LRDIMM. This baseline LRDIMM organization is used when the host operates in non-NDP mode.

To enable NDP, the authors propose adding simple cores to each data buffer chip. These cores are directly connected to a small set of DRAM chips. When NDP is engaged, i.e., when the host spawns threads on the data buffer chips, the interconnects assume new functionalities. A DRAM chip must now receive commands and addresses from its data buffer chip, not from the central chip. This is enabled by modifying the DRAM chips so that signals received on its pins can be interpreted as either data or commands/addresses with appropriate muxing and de-muxing. The authors show that dedicating some pins for commands/addresses in NDP mode is better for performance because it avoids the overheads of dynamic adaptation. Further, the short link between the data buffer chip and the DRAM chip in a baseline LRDIMM operates at a modest frequency because it only has to match the frequencey of the shared DDR4 channel. But in NDP mode, the interconnect between the DRAM chip and data buffer chip operates independent of the DDR4 channel; since this interconnect has a small length/load, it can operate at a higher frequency, referred to as a *gear-up* mode. This helps bridge the performance gap between a DIMM-based NDP architecture and an HMC-based NDP architecture.

An open question that remains: how should the DIMM be organized so it can switch between NDP and non-NDP modes, but without requiring a change to the DRAM chip and while offering high performance?

10.2   IN-SITU IMPLEMENTATIONS

The processor accesses data through a series of "pipes," with the pipes getting narrower near the processor. Each DRAM array has hundreds of bitlines that bring data from DRAM cells to their corresponding sense-amps. A small amount of that data is then shipped from the sense-amps to the DRAM chip's IO interface. The IO interface shares a narrow memory channel with several other DRAM chips. Thus, the DRAM arrays collectively have a very wide data pipe; the IO interfaces in DRAM chips collectively have a narrower data pipe; the memory channel feeding the processor has an even narrower data pipe.

NDP overcomes the last narrow pipe by placing the processor within the memory package. But it is unable to tap into the very wide data pipe within the DRAM arrays. To take advantage of the bandwidth available within the many array bitlines in the memory system, a few recent works have explored the concept of *in situ* processing. In essence, multiple cells in an array can exploit their shared bitline to perform specific operations within the array itself. This not only reduces overall data movement energy, it offers a much higher degree of parallelism. In situ processing is primarily enabled by new circuits and new memory cells. New non-volatile memory cells can be very helpful because they enable processing without losing the input operands, as is typically the case with destructive DRAM reads. We'll next discuss some of the major implementations. It is worth noting that such in-array computations are typically performed without access to error detection/correction codes; the reliability trade-off is usually not considered or not viewed as a liability in most literature in this area.

Computations in Resistive Memories

Some non-volatile memories, such as memristors, are implemented as crossbars, i.e., a grid of wires with a resistive cell at every wire intersection, shown in Figure 10.1. Such crossbars are not only useful as a dense memory unit, they can also be easily adapted to perform some computations on the resident data. As Figure 10.1 shows, when a vector of voltages is applied to the wordlines, the current through a bitline is a dot-product of the vector of voltages and the vector of conductances for the cells connected to that bitline. This basic operation simply relies on Kirchoff's Law to efficiently read operands, perform multiplications, and aggregate results. Since the input vector is fed to several bitlines, a number of dot-products are performed in parallel. In essence, the crossbar can perform a single-step vector-matrix multiplication, where the large matrix does not require any data movement. The catch here is that the analog values carried by the bitline currents have to be converted into digital values; the analog-to-digital converter (ADC), is a unit that consumes power and area, that grows significantly with ADC resolution.

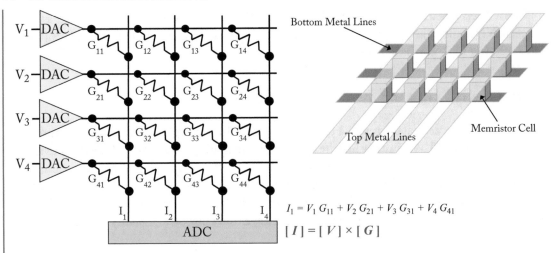

Figure 10.1: A dot-product computation with a memristor crossbar. The left figure shows a logical view of how Kirchoff's Law dictates the current in each bitline. The vector of currents, I, is the product of the vector of input voltages, V, and the matrix of conductances, G. DAC is digital-to-analog conversion, ADC is analog-to-digital conversion. The figure on the right shows a physical representation of the crossbar. The horizontal wires are on the bottom metal layer, the vertical wires are on the top metal layer, and the memristive material is sandwiched between the two layers at every overlapping point.

The ISAAC [191] and PRIME [192] architectures show how a crossbar and minimal ADC configurations can be exploited to improve the efficiency of deep neural networks by an order of magnitude. Both architectures reduce the ADC resolution by splitting a neuron computation in time and space, i.e., taking multiple cycles to feed a multi-bit input, spreading a weight across multiple cells, and spreading a neuron's inputs across multiple crossbars. PRIME also drops some unnecessary bits, while ISAAC encodes the data to promote small values. Follow-up work, PipeLayer [193], attempts to design a shallow pipeline geared for both inference and training.

Since analog units are vulnerable to noise, temperature, and other parasitics, these architectures also demand a degree of error tolerance. This has been tackled in a couple of ways. Recent work from IBM [194] combines an analog memristive compute unit with a traditional digital core; while the memristive unit does most of the computation to efficiently arrive at an approximate solution, light computation on the digital core helps refine the answer. Feinberg et al. [195] show how errors in analog computations can be detected and corrected with AN-codes. Weights are multiplied by a constant before placement in the crossbar; if an emerging dot-product is not a multiple of that constant, it indicates an error and some clues to help correct that error.

Feinberg et al. [196] also introduced a few techniques to efficiently map floating-point vector-matrix multiplication, vital for scientific computing workloads, to a crossbar. These techniques take advantage of locality in the exponent and eliminate ineffectual computations. Fujiki et al. [197] define additional operations for resistive crossbars that exploit bitline parallelism to support 13 SIMD instructions. They also develop a TensorFlow-based compilation framework so scientific applications can exploit these SIMD instructions, offloading as many as 87% of all instructions to memory.

Another body of work [198, 199] shows how memristor cells can be configured to take inputs from two cells, perform logic operations within the memristor array, and place the output in another memristor cell. This is an approach that was also later exploited within Compute Caches [200], although with SRAM-specific circuits. Another design, Pinatubo [201], modifies the sense-amplifier circuit to recognize more NVM resistance states, corresponding to the outputs of bitwise operations on two activated rows.

Finally, Guo et al. [202] showed how resistive cells can be leveraged to design a new TCAM cell that offers high density and low energy. TCAMs are used to find entries that match a key while allowing wildcards; they are widely used in networking applications. Their TCAM effectively uses 3 1T1R PCM cells per bit and improves TCAM density by 20×. The TCAM chips are placed on a DIMM with a DDR interface so they can also be used for conventional data storage. The subsequent AC-DIMM [203] improves the TCAM cell by using 2T1R STT-RAM and by co-locating the key and value. It also introduces programmable micro-controllers on the TCAM chip so that TCAM outputs can be processed by user-defined kernels.

In-SRAM Operations

Aga et al. [200] consider a form of near-data processing that moves computation into large SRAM caches on the processor chip itself; it is therefore targeted at reducing data movement between the caches and the core, while also leveraging the parallelism in thousands of cache bitlines. Recent work [204] has introduced circuits that allow a bitline to perform simple bit operations. For example, if a bitline is precharged to 1 and two wordlines are enabled, the bitline stays high only if both cells connected to that bitline are high. The output of this bitline therefore matches an AND function. Such bit-level operations can be composed to perform more complex computations; while this increases latency for a computation, this is offset by a large amount of parallelism. Some compiler support is required to promote operand locality, i.e., operands that are aligned in subarrays and amenable to in-cache operations. Aga et al. show that a Compute Cache can be designed to support the following vector computations: copy, search, compare, logic (and, or, xor, not). Other, more complex computations can be performed in ALUs beside the subarray. All of this adds an area overhead of 8% per subarray. The supported computations are useful for a number of workloads: text processing, databases, operating systems, cryptography, and bioinformatics. In follow-up work, the authors also show how this approach can accelerate neural network computations [205].

PROMISE is another architecture that leverages computations in SRAM [206] and targets machine learning workloads. Unlike the Compute Cache, PROMISE uses a mix of analog and digital circuits, with the following SRAM access at the heart of the design. A w-bit word is mapped to multiple SRAM cells sharing the same bitline; each of the wordlines is then activated simultaneously, with the activation pulse width proportional to the bit position; the resulting bitline voltage drop is an analog representation of the word. The authors introduce various operators and create a compiler to map machine learning algorithms in a high level language to PROMISE hardware.

The NAND-net architecture of Kim et al. [207] focuses on binary neural networks and shows how the computation can be transformed to leverage operators (NANDs and popcounts) that are easier to implement in-SRAM and in-DRAM.

In-DRAM Operations

The DRISA architecture [208] modifies a DRAM chip to have compute capability. The authors acknowledge that this is likely not a cost-effective way to produce dense memory chips, but it may be an effective strategy to produce accelerator chips. Because of destructive reads in DRAM, the design space includes the following: (i) create a new cell, e.g., 3T1C, such that reads are not destructive, and perform basic logic operations in the bitline by activating multiple rows (similar to Compute Cache above and introduced earlier by Seshadri et al. [209]); (ii) retain the standard 1T1C DRAM cell, but copy the operand rows [86] before performing any bitline computations; (iii) and retain the 1T1C cell, but add a latch and ALU near the sense amplifier so that two operands can be read and operated upon. Li et al. carry out an exploration of the design space to identify that the last of the above options is likely the sweet spot if the ALU supports a limited number of simple operations. This is important because the DRAM process is highly inefficient for implementing logic. The authors show the effectiveness of these in-DRAM operations for deep neural network inference. In follow-up work [210], they note that multiplications require many cycles because of the bit-serial nature of DRISA; this is alleviated with a stochastic computing approach.

A body of work by Seshadri et al. also studies the effectiveness of operations within DRAM chips, such as bulk copies [86] and bitline AND/OR [209, 211]. The Ambit architecture [211] focuses on minimal changes to the DRAM chip (estimated area overhead of under 1%) and DDR interface. By adding support for highly parallel bitwise operations, they show improvements for database and web search applications. Seshadri et al. observe that when three DRAM rows are simultaneously activated, each sense-amplifier detects the majority value in the three cells connected to that bitline. By setting one of the rows to either all-0 or all-1, the result is the AND or OR of the other two rows. The result can also be inverted by leveraging a NOT gate within the sense-amplifier. The DRAM decoding logic can also be simplified by fixing the operand rows for the AND/OR operations. Operands must therefore be copied to/from these operand rows before/after the operation; this is anyway necessary given the destructive nature of the triple row activation.

Automata Processor [212]

Micron created the Automata Processor (AP) as an experimental testbed. The AP is a chip that combines storage and compute, uses a non-von Neumann model, and resembles the Multiple Instruction Single Data model in Flynn's taxonomy. The AP is composed of several DRAM mats, with each column connected to some simple logic, and a hierarchical routing network to carry outputs. Non-deterministic finite state automatons (NFAs) can be mapped to the AP; as each input symbol arrives, several states respond in parallel. A DRAM mat has 256 rows; each column represents a state; when an 8-bit input symbol arrives, it reads the corresponding row out of the mat; if a state has a 1 in that row, a match is indicated. The logic for that state/column then processes the match; this may involve a latch that stores whether the state is active or not, a counter, and some Boolean logic. A hierarchical routing element then carries the output of this state to other relevant entities. The AP is highly configurable. The user programs the DRAM rows, the initial states, the counters, logic, routing, etc. The first-generation boards consisted of 32 chips, each consuming 4 W, and collectively capable of storing 1.5 million states. This offers very high levels of parallelism for applications performing exact or inexact matches [213–215], i.e., regular expression matching, DNA analysis, data mining, text search, etc.

10.3   PROGRAMMING MODELS AND APPLICATIONS

A key challenge in realizing NDP is the software burden. What are the best programming abstractions for offloading computations to an NDP unit, how does the programmer/runtime/hardware identify when NDP is useful, and which classes of applications are likely to benefit the most? The many papers discussed so far have already shown NDP benefits for a range of applications: in-memory MapReduce, pattern matching, database/web search, to name a few. In this section, we'll briefly go over a few more workloads and a few new approaches to ease the programming burden.

10.3.1   PROGRAMMING APPROACHES

A large fraction of NDP papers focus on the underlying architecture, while assuming that threads are spawned as in traditional systems with or without coherent memory (GPUs, accelerators, multi-cores). A few, such as the AP discussed earlier, introduce altogether new programming abstractions. A number of works also assume distributed system runtime frameworks, such as MapReduce, to manage tasks. Below, we discuss approaches to seamlessly leverage existing distributed system frameworks, as well as approaches that modify the programming model for NDP.

MCN, Alian et al. [178]

The memory channel network (MCN) considers a novel approach to overcome the software/runtime problem for NDP. It implements device drivers on NDP cores that provide the illusion of an ethernet link. A large collection of NDP cores therefore appears as an ethernet-connected cluster; each NDP core appears as an independent system connected to its local memory. Applications designed for distributed systems can therefore run unmodified on MCN. The authors also create a prototype with an FPGA on a buffered DIMM. A number of optimizations are introduced to improve bandwidth and reduce latency. On the software front, some of the TCP/IP stack features are customized for MCN requirements, while on the hardware front, a DDR-based interrupt signal and a hardware DMA are leveraged. The authors demonstrate that the overall higher memory bandwidth enables a 4.5× speedup over conventional ethernet-connected nodes.

PEI, Ahn et al. [216]

Ahn et al. introduce an approach where the program is untouched, but some instructions are PIM-enabled, i.e., they may be offloaded to a memory device capable of executing that instruction. For example, if a single variable is being updated in a traditional system, the entire 64-byte cache line is fetched from memory, a few bytes are modified, and the entire line is then written back to memory. Instead, to save memory bandwidth, a few bytes (the increment) can be sent to memory and the increment can be performed on the memory device. PIM-enabled instructions (PEI) are therefore an attempt to exploit fine-grained opportunities for low data movement. Of course, the PEI should be offloaded to wherever the instruction operands currently reside, either a compute-enabled cache or a compute-enabled memory device. Ahn et al. introduce a locality monitor to help decide if and where the PEI should be offloaded.

GraphPIM, Nai et al. [217]

Nai et al. extend the approach of Ahn et al. with a focus on graph workloads. They analyze these workloads and identify that atomic graph operations account for most irregular memory accesses and are easily identifiable as appropriate candidates for PIM offload. They also take advantage of 18 atomic operations identified as PIM operations in the HMC 2.0 specification. The authors introduce an offload unit on the processor that first marks some regions of memory as uncacheable and then offloads atomic operations on those regions to the HMC device.

Active-Routing, Huang et al. [218]

Huang et al. note that in many applications, once computations are offloaded to HMC devices, the results also have to be aggregated. To reduce the bandwidth overhead on the processor, the aggregation is performed within the HMC network. A dynamic tree is constructed during the offload process; once a node performs its compute, it aggregates its result with those of its children before returning the result to the parent node.

TOM, Hsieh et al. [219]

Hsieh et al. target a system with GPU cores scattered across HMC-like devices. They introduce compiler and OS techniques to not only determine code that should be offloaded to near-data GPUs, but also improve data affinity by predicting and placing pages touched by offloaded code to the corresponding memory stack. Follow-up work [220] relieves the NDP stack of its memory management duties and factors cache locality into the decision-making.

10.3.2 BROADENING THE SCOPE OF NDP WITH ADDITIONAL WORKLOADS

Much of the early NDP work focused on applications with clearly large amounts of parallelism, often expressible with frameworks like MapReduce. In recent years, benefits have also been demonstrated for other workloads, including the impact of 3D stacked NDP on low-latency virtual reality [221], the use of in-DRAM operators for stochastic computing [210], and NIC support in 3D stacked memory to accelerate Memcached servers [222].

For a summary of recent work that moves compute closer to SSD, we point the reader to a blog post by Blanas [223]. In that post, the author makes the argument that for many big-data workloads, SSD storage is likely the sweet spot; placing compute in the SSD controller helps exploit higher bandwidth. The available bandwidth is even higher when compute is moved closer to the NAND dies, but at that point, management of frequent errors must also be considered.

Below, we discuss some of the other relevant workload classes that have been accelerated with NDP.

Deep Networks

Not surprisingly, deep neural networks (DNNs) is one of these relevant workloads. The fully-connected layers of DNNs are typically memory bound, while convolutional layers require moving large amounts of data between caches and compute. NDP has therefore been a vital element in DNN architectures, starting from DaDianNao [224] that places each neural functional unit adjacent to an eDRAM bank, to Neurocube [225] and Tetris [226] that place neural units under 3D-stacked DRAM, to more recent work that targets DNN training [227, 228] and generative adversarial network training [229].

Graph Processing

A body of work has also tried to accelerate graph workloads. We have already discussed atomic operations offloaded by the host processor in GraphPIM [217] above. GraphR [230] uses analog computations with resistive crossbars; given the inherent imprecision, the authors focus on graph algorithms that are either iterative or computing approximate statistics. Efficiency is achieved by converting graphs into matrix operations that leverage the dot-product capabilities of resistive crossbars. Tesseract [231] is a system that distributes a graph across several HMC-like devices with a vertex-centric model; operations on graph vertices are performed on the HMC device; communication is required every time an adjacent vertex is mapped to a different HMC. The authors introduce non-blocking message-passing primitives and prefetching to hide these communication costs. The messages include prefetch hints so that a non-preemptable remote function call, e.g., an atomic memory update, is invoked only after relevant data has been prefetched into buffers. In follow-up work, Zhang et al. [232] observe that inter-HMC communication is a significant bottleneck. They introduce GraphP that uses "source-cut" partitioning that maps a vertex and all its edges to the same HMC; the result is that edges/vertices are replicated and the nature of communication shifts to replica updates. This co-design of the programming model and the graph partition helps reduce the communication bottleneck.

Mondrian Data Engine, Drumond et al. [233]

The Mondrian Data Engine [233] makes the case that many analytics algorithms, e.g., many MapReduce workloads, have data shuffle phases that involve random accesses and a loss in row buffer locality. The authors observe that data being received during a shuffle can often be steered to a single DRAM row because permuting the data has no impact on correctness. By forcing the algorithm to have more streaming access patterns, the hardware can also be kept simple; a modest SIMD core can efficiently utilize the available memory bandwidth, out-performing prior NDP systems by about 5×.

Consumer Workloads, Boroumand et al. [234]

The work of Boroumand et al. [234] explores the potential of NDP for Google workloads (Chrome, TensorFlow Mobile) on consumer devices. They quantify that 63% of system energy can be attributed to data movement between the processor and memory. Similar to prior work, NDP is implemented by adding an in-order simple core (e.g., ARM Cortex-R8) to each vault of an HMC-like device. The authors argue that such cores impose a small (less than 10%) area overhead, although the impact of using 3D-stacked memory on the cost of the consumer device is less evident. In each workload, they identify specific functions like memset, memcopy, and other bitwise operations that can be performed on the HMC. This causes a 2× reduction in execution time and system energy.

10.4   DISCUSSION

For some of the chapters in this book, we have ended with a conjecture that the low-hanging fruit has been picked. That is not the case with NDP, where I believe we have barely scratched the surface. Moving forward, more killer applications will be discovered, more commercial offerings will support NDP, software will be overhauled to exploit and expose more NDP, and we will find new ways to integrate more operators into SRAM/DRAM/NVM—this is an extremely fertile area for young researchers looking for a multitude of open problems. A significant fraction of the work in this area focused on devices resembling Micron's HMC; with Micron moving away from the HMC, the memory landscape is less clear. While NDP ideas await a commercially viable 3D-stacked platform, we need to more aggressively pursue NDP incarnations that exploit 2.5D stacking, other novel interconnects, and in/near-array compute.

CHAPTER 11

Memory Security

In a system that provides high security and privacy, both computation and memory must be protected. The memory system is vulnerable to several attacks, partly because memory modules are easy to replace with malicious memory modules that can observe program behavior and disrupt its execution.

Obviously, if an application cares about security and privacy, it cannot afford to place its data in memory in plaintext form. So let's first assume that the system has basic primitives in place: (i) data emerging out of the processor has been encrypted; and (ii) the hardware and OS ensure authentication, i.e., an application can read and decrypt data from memory only if it has permissions to do so.

In spite of these basic primitives, attacks can be launched to steal secrets through side channels and to disrupt execution. In the security literature, a side channel is a flow of information from a victim application to an attacker, typically by indirect means. Side channels can take many forms—the attacker can measure electromagnetic emanations, power draw, memory activity, patterns of memory accesses, branch activity within the CPU, etc. There are three main types of memory vulnerabilities: (i) memory timing channels, where an attacker application can learn secrets of other co-scheduled applications by measuring its memory latencies; (ii) memory integrity violation, where an attacker can modify the data being returned to the processor; and (iii) memory access pattern leakage, where an attacker can learn secrets by observing the addresses being touched by an application.

Several solutions have been developed to defend against such attacks. Unfortunately, many of these solutions have very high overheads, lowering overall performance by orders of magnitude in some cases. In the last two decades, some of these solutions have advanced from crazy academic exercises to commercial reality. In the next three sections, we will examine each of the memory vulnerabilities and state-of-the-art solutions. We'll finally discuss the impact of emerging active memories (with logic capabilities) and other potential attacks on the memory system.

11.1 MEMORY TIMING CHANNELS

Attack Scenarios

A timing channel is a special case of a side channel. In a typical timing channel, the attacker, while running on a shared computer, takes a timing measurement for some task. The latencies experienced for that task can indicate the activity levels for other tasks (victims) running on that shared computer. These activity levels can sometimes reveal the victim's secrets. Consider two important example attacks.

In an RSA decryption algorithm, certain operations that can yield cache misses are performed only when a "1" is encountered in the private key. If an attacker keeps accessing memory and observes a higher than usual latency for memory access, they can conclude that the victim is also issuing many requests to memory, i.e., it may have a private key that has a large number of 1s [235]. This helps the attacker narrow down the search space and carry out a brute-force search to identify the private key. This may appear like science fiction and most layman reactions to this attack are laced with incredulity. Indeed, this is not a sure-fire attack—for most such hacks, a persistent attacker must make several failed attempts before they get lucky. And such attacks have been demonstrated in the field—Ristenpart et al. [236] were able to exploit cache timing channels to crack passwords on an Amazon EC2 infrastructure. The work of Pessl et al. [237] shows how memory address mapping in a system can be reverse-engineered, and how this helps an attacker exploit row buffer hits and misses to construct a timing channel.

Now consider a second attack that is even easier to pull off. Imagine an electronic health record infrastructure executing on the cloud. The infrastructure may use some third-party software, e.g., a document reader. Even if all the data on the cloud is in encrypted form, the document reader will have access to unencrypted data. But assuming that the cloud infrastructure uses all the right firewalls, the document reader cannot release the private data to the outside world. This is where a timing side channel can be used by the document reader to leak secrets to a cohort application running on the same server. In essence, the untrusted third-party software is a trojan. The side channel that is set up between the trojan and its accomplice is almost impossible to detect and is referred to as a *covert channel*. For example, the trojan can issue a burst of memory accesses to represent a 1; the accomplice is constantly accessing memory; every time the accomplice sees a spike in its memory latency, it detects the high memory contention and recognizes the transmission of a 1. Hunger et al. [238] demonstrate that covert channels with bandwidths of as high as 100 Kbps can be established between two threads on a server. Pessl et al. [237] advance that effort by using row buffer interference to construct a 2 Mbps covert channel.

When trying to build a defense against timing channels, it is helpful to confirm that the defense can defeat establishment of a covert channel. If the defense can overcome a covert channel where both trojan and receiver are actively colluding, it can certainly defeat the first scenario above where the attacker gets no help from the victim.

Temporal Partitioning

Wang et al. [235] were the first to design a memory controller that eliminates timing channels within a shared memory controller. For the rest of the discussion in this section, we will assume that multiple independent virtual machines (VMs) are scheduled on a cloud server, and we will only discuss operations in a single memory channel shared by all the VMs. Among these co-scheduled VMs is an attacker that constantly accesses memory and monitors latencies to estimate the memory intensity in other VMs.

In the *Temporal Partitioning* (TP) design of Wang et al., each VM receives a *turn* to use the memory controller, and turns are allocated to each VM in round-robin order (or any pre-determined input-independent order). Each turn has a fixed *turn length*. When a VM begins its turn, it should be free to issue any memory requests it wants. If any of these requests is disallowed because of DRAM timing constraints, it leaks information about what the previous VM might have done in its turn. Therefore, a VM must complete all its operations before its turn ends, thus leaving no trace of its activity within the memory system. Accordingly, each turn ends with a *dead time*; during this dead time, the VM cannot issue any new operations because they are not guaranteed to finish before the end of the turn. The dead time essentially isolates each VM's execution and eliminates timing channels.

While a non-secure baseline is free to issue memory requests from any VM at any time, TP allows only one VM to issue requests at a time and further prevents that VM from issuing requests during its dead time. It therefore introduces a significant performance penalty, relative to the non-secure baseline. One of the keys to improving performance is to reduce the length of the dead time. Based on DRAM timing parameters, this length is the time taken to perform a write operation in a bank, followed by a precharge of that bank, which for some common DRAMs is between 40–50 DRAM cycles. Wang et al. also observe that if two consecutive VMs accessed different banks, then the dead time can be lowered, i.e., one VM could be wrapping up its operations in one bank while another VM could start its turn and access a different bank. To enable this technique, the memory banks must be partitioned among the VMs such that no bank is shared by multiple VMs. With such *bank partitioning*, the dead time can be reduced to 15–20 DRAM cycles (see example in Figure 11.2a).

Fixed Service

Shafiee et al. [239] build on the work of Wang et al. with a generalized mathematical framework to create a non-interfering pipeline of memory requests from different VMs. In essence, each VM injects a memory request in round-robin fashion; a dummy operation is injected if a VM has no pending request. These memory requests are separated by a fixed gap such that they flow through the memory system without competing with each other for resources and without violating any DRAM timing constraints. By solving a system of equations that define these timing constraints, we can compute the minimum value for this fixed gap.

Shafiee et al. show that for their assumed DRAMs, the gap between two memory requests can be as little as seven cycles if they are destined to different ranks. Similar to the work of Wang et al., they show that the gap is 15 cycles if the two requests are destined to different banks, and the gap is 43 cycles if the two requests are to the same bank. Thus, performance is highest if two VMs are guaranteed to not access the same rank (*rank partitioning*).

To improve performance in the cases that use bank partitioning and no partitioning, Shafiee et al. introduce two new techniques. In the bank partitioning case, the 15-cycle gap is incurred because in the worst case, a VM may inject a write, and the next VM may inject a read. In all other cases, the gap can be as low as six cycles. Therefore, requests are processed in batches, with each VM contributing one request to the batch. The memory controller first processes the reads in that batch, followed by the writes (all with six-cycle gaps). A 15-cycle gap is introduced only before starting the reads in the next batch. To not reveal the read/write behavior of each VM, all the read values are sent from the memory controller to the cores at the end of the batch.

The second new technique of Shafiee et al. is targeted at the case where we make no assumptions about the OS page allocation policy, i.e., there is no bank or rank partitioning. The gap between requests from two VMs is 43 cycles in this case because both VMs may access the same bank. Therefore, the memory controller introduces additional constraints to prevent this. The first VM is only allowed to access banks that are multiples of three; the second VM is only allowed to access banks that are multiples of three plus one; the third VM is only allowed to access banks that are multiples of three plus two, and so on. This guarantees that consecutive requests do not go to the same bank, i.e., this emulates a bank-partitioned model. With this *triple alternation* technique (see example in Figure 11.2b), the gap between consecutive requests is only 15 cycles. If a VM does not have sufficient memory-level parallelism, it may not have a pending memory request that fulfils the memory controller criteria, thus increasing the number of dummy accesses.

Figure 11.1 shows the performance of each of these discussed models, relative to a baseline that allows timing channels. By far, rank-partitioning is the best (only a 26% slowdown), while a system with no partitioning incurs a 60% slowdown, even with the triple alternation technique.

Dynamic Scheduling

Wang et al. [240] continued this progression in timing-channel-free memory controllers with a dynamic scheduling technique to extract higher performance without any rank or bank partitioning support from the OS.

In their technique, each VM receives a somewhat large turn so it can issue multiple memory accesses. Assume that this turn length is 43 cycles. Within this turn, a VM can issue six memory requests—two from one rank, two from a second rank, and two from a third rank (see example in Figure 11.2c). All six requests are sent to different banks. A conflict-free schedule is set up by first issuing the first request to the three ranks, followed by the second request to the three ranks. Following the rules described earlier, consecutive requests to different ranks can be issued with a gap of only seven cycles. Requests to different banks in a rank are being separated by 21 cycles, which also fulfills the previously discussed rules.

Depending on the ranks and banks selected by the previous VM, some of the requests of the current VM may have to be scheduled toward the tail end of the turn to ensure a sufficient gap between accesses to the same bank from different VMs. Because the banks being accessed in a turn are unique, we are guaranteed to find a correct conflict-free schedule for every turn, in spite of the constraints imposed by the schedule in the previous turn. Thus, a given VM is able

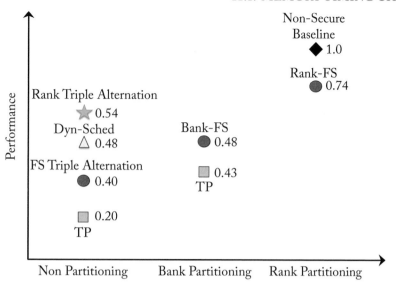

Figure 11.1: Representative performance of different timing-channel-free memory controller policies with varying OS support for VM page allocation, relative to a non-secure baseline.

to issue any requests it desires in a turn (to three ranks) regardless of what other VMs are doing, hence the *dynamic scheduling* moniker. The schedule within a turn may reveal the banks touched by the previous VM. To eliminate this leakage, all read requests are returned to the CPU at the end of a turn; this ensures that the VM is unaware of the schedule that is selected by the memory controller.

Vuong et al. [241] extended the dynamic scheduling approach of Wang et al. by first observing that most workloads do not have sufficient memory-level parallelism to issue six memory requests in every turn. The turn length is therefore shortened to 14 cycles and a VM can issue two requests to two ranks in its turn. Rank triple alternation is employed, i.e., in the first turn, the VM accesses ranks that are multiples of 3; in the next turn, the next VM accesses ranks of the form $3N + 1$, and so on. This rank triple alternation technique is the state-of-the-art leakage-free scheduler that does not require OS support.

Figure 11.2 captures the differences between the various timing-channel-free memory controllers described in this section. Figure 11.1 depicts the performance of this design space.

Performance Trade-Offs

The best timing-channel-free memory controllers to date continue to incur 2× slowdowns (assuming no rank or bank partitioning by the OS). Therefore, more advances are required. To bring down this overhead in certain scenarios, Wang et al. explored two other strategies.

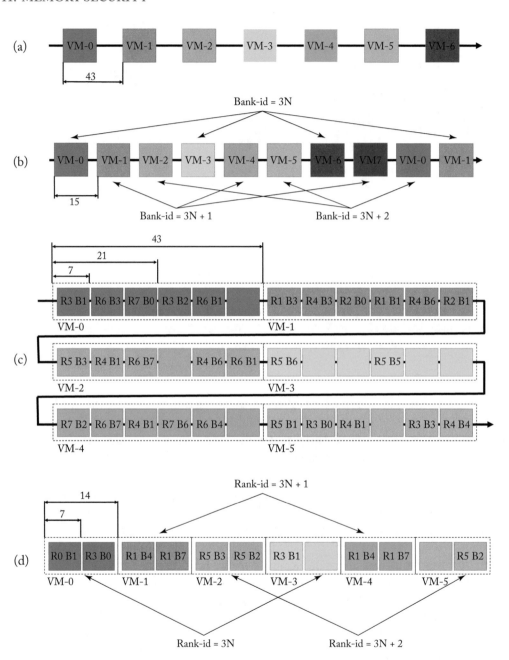

Figure 11.2: (a) A comparison of scheduling policies in Temporal Partitioning, (b) FS:Bank Triple Alternation, (c) Dynamic Scheduling, and (d) Rank Triple Alternation.

In the first strategy [242], they assume that VMs with different security requirements may share a single server. Some sensitive VMs cannot tolerate information leakage to other VMs, while some non-sensitive VMs can afford to leak information to some other VMs. These security requirements can be represented as a lattice structure to indicate acceptable information flows among VMs. They then design a *lattice priority scheduler* to extract higher performance while allowing benign information leakage. Consider a highly sensitive VM H and a low sensitive VM L; L may leak information to H, but not the other way around. First, L is allowed to use as many turns as its maximum allocation, i.e., L is going to receive nearly peak performance. After this is done, H receives any remaining turns, i.e., H receives at least its guaranteed service and in some cases, it receives even higher service when L has low memory demands. Note that L is unaware of the turns that were allocated by the memory controller to H. Second, L is allowed to issue memory requests during its turn's dead time. This offers higher performance to L at the expense of H. By introducing both policies, a significant improvement is observed for low sensitive VM L. For VM H, there is a positive effect from the first policy and a negative effect from the second; the net performance impact is typically positive.

In their second strategy, Wang et al. [240] trade off security and performance. They introduce a technique to improve performance, while allowing a bounded amount of information leakage. Every VM is assigned an *expected response time* (ER), which is a conservative estimate for the memory latencies experienced by this VM. The memory controller then forces every memory access from this VM to return at this pre-defined ER. Thus, the memory latencies will be a little higher than the memory latencies experienced by a non-secure baseline. The hope is that a slightly conservative ER will succeed most of the time, and the imposed performance penalty will be smaller than the 2× penalty imposed by other timing-channel-free memory controllers. But for a sufficiently small ER, there will be occasions where the memory access cannot be serviced within ER. When this happens, the memory access is returned at a larger pre-defined response time. Every such *ER violation* leaks information. To bound this information leakage, the memory controller only allows a fixed number of such violations in every epoch. Once this limit is reached, the memory controller has to default to a guaranteed timing-channel-free memory controller for the rest of the epoch. A VM now has multiple knobs at its disposal to tune the performance/security trade-off: epoch length, violation count, and ER.

11.2   OBLIVIOUS RAM (ORAM)

Attack Scenarios

Even if all data in encrypted, a malicious agent can use various tricks to observe how a program is bouncing across various data structures or across various parts of the code. The information leak from this side channel can then be used by the attacker to perhaps reverse engineer the algorithm being used or the inputs that may be fed to the victim program [243]. For example, the attacker can estimate that given the data structures being touched, a genome being sequenced is vulnerable to say Alzheimer's, thus compromising patient privacy.

One may argue that randomizing the page layout and frequently shuffling pages should be enough to confuse the attacker. But "confusion" or "obfuscation" is usually not enough to thwart a persistent attacker with vast resources. Even with such shuffling of data structures, an attacker can determine reuse distance, read/write ratio, temporal/spatial locality, distributions of page accesses, etc. For guaranteed privacy, the system must offer the property of "indistinguishability"— given two different inputs to a program and the resulting memory access patterns, it should be impossible to correlate the input and the memory access pattern. For decades, the ORAM construct has been the gold standard for providing the indistinguishability property [244]. In recent years, the Path ORAM algorithm [245] has emerged as the most popular.

But first, let's discuss how an attacker gets a window into an application's access pattern. A malicious OS can periodically flush the TLB or swap out application pages; when the victim application page faults or asks for TLB refills, the OS identifies the pages being touched [246–248]. This is information leakage at the coarse granularity of pages. A second attack may be possible through a compromised DMA which can capture images of the memory [249]; by comparing two consecutive images, the attacker can identify which lines were modified. This attack can however not identify blocks that were read. Both of these attacks can be launched remotely, i.e., the attacker need not have physical access to the hardware. In the third attack scenario, the attacker has physical access to the hardware. With a specialized DIMM or with a logic analyzer, the attacker can monitor all transactions on the memory bus. This is the worst-case attack typically assumed in most ORAM literature.

A key assumption in ORAM literature is that data blocks are encrypted, but memory addresses are not. Thus, by monitoring the memory bus, the attacker learns the memory access pattern. This is a valid assumption for most main memory systems. Most main memory systems are constructed with chips that are trying to eke out maximum storage density. Memory vendors therefore have not added extra logic on their DRAM chips to perform encryption/decryption of data or addresses. The memory chip must therefore receive an address in plaintext and it simply reads/writes the data block in that address. However, this assumption may change in the future and we will discuss those scenarios in Section 11.4.

Path ORAM

At CCS 2013, Stefanov et al. [245] introduced the Path-ORAM algorithm and demonstrated an implementation of Path-ORAM on an FPGA-based platform (Phantom [250]). While implementable, the design has a bandwidth requirement that is more than two orders of magnitude higher than that of the native application without ORAM support.

Path-ORAM organizes the memory into *buckets* that are composed of Z data blocks and an additional block for metadata. These buckets are organized into a logical tree structure. Every data block is associated with one leaf vertex; this association is tracked in a large position map (PosMap) table at the memory controller. Figure 11.3 shows an example of a requested block A that is mapped to leaf vertex 0, and that is resident in bucket P. When block A is requested by the processor, a look-up of the PosMap yields the leaf vertex 0. All buckets in the path between

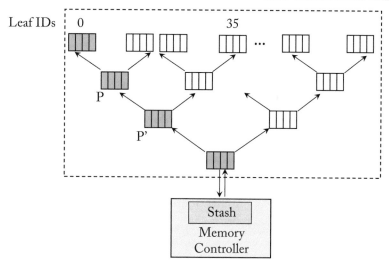

Figure 11.3: Path ORAM: a block is moved from bucket P to bucket P'. The block was previously mapped to leaf ID 0, and is now mapped to leaf ID 35.

the root of the tree and the leaf vertex (colored in blue) are fetched from memory. The fetched blocks are placed in a *stash*. The block is then associated with another random leaf vertex, 35 in this example. Next, a number of blocks in the stash are re-encrypted and written back to the vertices between the root and the old leaf vertex. The stash is scanned to find as many blocks for the write-back that can occupy positions on that path. In this example, block A is placed in a bucket P' that appears in the path from the root to its new leaf vertex 35. If candidate blocks are not found, dummy blocks are written back. About half the memory capacity is occupied by dummy blocks. Also note that the read operation brought in a number of dummies that are discarded and not placed in the stash.

The key here is that every data block access fetches and updates a random path in the tree. During this process, the requested block has changed its leaf vertex in an untraceable manner. Thus, regardless of the data access pattern within the program, it manifests as random path fetches/updates in memory.

Earlier studies, including Phantom [250], used large block sizes to reduce PosMap size, thus paying a very high memory bandwidth penalty. Recent studies have used conventional 64-byte block sizes to reduce this penalty. The resulting large PosMap can itself be stored in memory as a recursively accessed ORAM [251].

Path ORAM exhibits other key trade-offs. A large bucket size, Z, increases the memory bandwidth penalty, but leads to fewer stash overflows. Fletcher et al. [251] show that $Z = 4$ is a sweet spot.

The overhead introduced by ORAM is a function of the number of required recursive look-ups, the depth of the ORAM tree, the size of the bucket, and the size of the block. More specifically, the bandwidth overhead is expressed as $R \times (Z + 1) \times L \times 2$, where $R = 1.55$ is the average number of recursive accesses [251], $Z = 4$ is the bucket size, and $L = 28$ is the depth of the ORAM tree for a 64 GB memory. Thus, every memory access by the program translates into 217 memory reads and 217 memory writes on average. Ouch!

This two orders of magnitude bandwidth overhead for ORAM was always assumed to be too high to be practical. But just as integrity verification (Section 11.3) has now become practical, the hope is that continued advances will eventually make ORAM overheads tolerable.

ORAM and DRAM Co-Design

In this section, we will discuss a number of innovations where either the ORAM has been tailored to lower its pressure on the memory system, or the memory system architecture has been tailored for ORAM requirements.

Phantom [250] was one of the first demonstrations of ORAM's practicality, using a Convey HC-2ex FPGA-based platform. It reports a slowdown of only 1.2–6× for various workloads. However, this is possible because the platform accommodates 16 memory channels and 1 K banks. The high parallelism in this hardware system cannot be exploited by a non-secure baseline application, but it is perfect for a program augmented with ORAM. Phantom doesn't help Path ORAM defy gravity—if a baseline non-secure system utilizes 100% of its memory bandwidth, adding Path ORAM to it will slow the system by two orders of magnitude.

Ren et al. [252] introduce two ORAM modifications to exploit memory system effects. First, to exploit spatial locality without information leak and without increasing block size, they introduce the notion of a super block. A super block is composed of multiple independent blocks which are all assigned the same leaf vertex. Thus, bringing a tree path into the stash has the advantage of prefetching other relevant blocks in a super block. Ren et al. also partition the ORAM tree into sub-trees, with each sub-tree containing a small number of levels; for example, a sub-tree with three levels is made up of a root node (bucket), its children, and its grandchildren. A sub-tree is mapped to all memory channels, occupying one row buffer per channel. The multiple blocks fetched from a sub-tree thus enjoy a combination of row buffer locality and channel-level parallelism.

In the Fork Path architecture [253], Zhang et al. observe that consecutive ORAM requests access random paths, but any two paths have overlapping buckets. Those buckets need not be updated by the first request, only to be read and updated immediately by the second request. Those overlapping buckets are therefore not written to by the first ORAM request and not read by the second ORAM request; they are retained in the processor stash. This avoids several memory accesses for the top regions of the ORAM tree. Zhang et al. combine this path merging with a greedy scheduling policy that looks at the queue of pending ORAM requests and selects a next request that maximizes path overlap with the previous request.

The Secure DIMM architecture [254] modifies the memory system and the ORAM protocol. Each Secure DIMM (or SDIMM) is equipped with a buffer chip (similar to the buffer chip on an LRDIMM [15]) that is capable of exchanging encrypted messages with the processor. Since the buses on an SDIMM are exposed to an attacker, the buffer chip requires an ORAM-like protocol to read/write data on the SDIMM. Thus, the internal buses on an SDIMM bear the brunt of ORAM bandwidth overheads, while the processor is sheltered from most ORAM overheads. The controllers on the processor and the SDIMM buffer chips thus form a distributed *trusted computing base*. Shafiee et al. [254] introduce two distributed ORAM protocols that leverage SDIMMs to reduce latency and/or bandwidth for ORAM transactions.

The Relaxed Hierarchical ORAM [255] uses a two-level ORAM and reduces the overhead for the first level by (among other techniques) packing most metadata into space typically used for ECC. Zhang et al. [256] replace dummy blocks in ORAM with duplicate versions of blocks in the path; latency is reduced because the processor can proceed with the first version that is returned to the memory controller.

11.3 MEMORY INTEGRITY

Attack Scenarios

The third memory attack we'll cover is not a side channel. It is an attack where an application's execution is disrupted by feeding it incorrect values.

When the application writes data into a specific location in the memory system, it expects to receive exactly that same data when it subsequently reads from that location. A memory system that fulfils this expectation is said to provide *integrity* or *freshness* guarantees. A baseline memory system does not provide this guarantee as it is vulnerable to two kinds of attacks.

The first is a hardware attack where the attacker has physical access to the hardware. The attacker can install a custom DIMM with a buffer chip, similar in style to an LRDIMM [15]. The buffer chip can respond to requests with incorrect data.

The second attack is software based and relies on a compromised OS. In a baseline system, the OS has sufficient privileges to modify the contents of any page in memory, thus disrupting inputs to a victim application.

To protect from such disruptions, the processor hardware can provide some basic primitives. Every block of data (say a 64-byte cache line) can be associated with a hash—a *Message Authentication Code* (MAC)—that is also stored in memory. The MAC is computed by processor hardware using a one-way function with a private key. When data is read from memory, the processor re-computes the MAC and confirms that it matches the MAC received from memory. If the attacker tries to fabricate their own data block, they would also have to produce a correct MAC, which is intractable.

Of course, this basic primitive can be easily fooled. The attacker can simply return a data block and MAC corresponding to a different memory location in what is called a *splicing* or relocation attack. This can be thwarted by including the memory address in the MAC function.

We thus have a basic solution that only requires storing/retrieving a MAC per data block; this solution places as much burden on the memory system as an ECC solution, i.e., it is relatively inexpensive and serves as a first line of defense against an integrity violation.

This solution is, however, vulnerable to a *replay attack*. In a replay attack, the attacker returns an earlier version of a block/MAC that may have been written to that same address. A defense against a replay attack is what makes an integrity guarantee expensive. We next discuss the state-of-the-art defenses.

Based on the discussion above, there are two more ways to classify an integrity attack—one in which an attacker is (perhaps randomly) trying to modify data and one in which an attacker is very precisely trying to return data that can fool system defenses. The former attack is easy to pull off, maybe even with software that implements a Rowhammer attack [257], while the latter is harder. Correspondingly, the former attack can be thwarted with simple solutions, while the latter requires the more complex mechanisms that we will discuss shortly.

It is worth highlighting that integrity guarantees were once deemed too expensive to be commercially viable. Thanks to decades of research, that is no longer the case—many commercial designs, including Intel SGX, do provide integrity guarantees. This also lends credibility to the attack scenarios discussed above. However, as discussed later, the overheads of integrity in Intel SGX and other systems are non-trivial and require continued research.

Merkle Trees

A Merkle Tree is an auxiliary data structure stored in memory that defends against a replay attack. The MACs of all data blocks form the leaves of the tree. Every non-leaf node in the tree stores a hash of its child nodes. The root of the tree is stored on the processor and cannot be modified by the attacker. When a block and MAC are read by the processor, it also reads all the ancestors of that MAC in the Merkle Tree. The hash in every ancestor is confirmed by reading its child nodes and re-computing the hash of the child nodes. If an attacker tries to fake any block in this process, at least the hash in the root will not be confirmed with a very high probability.

As is evident, this integrity verification can be a very expensive process as it requires reading several nodes of the Merkle Tree. One way to lower the burden is to pack all the children of a node into a single memory block. If we assume that every MAC and every hash is 64 bits in size, we can pack 8 MACs or hashes into every 64-byte memory block. This means that the Merkle Tree has an arity of eight. It also means that one verification operation requires us to fetch one memory block from each level of the Merkle Tree. A 64-GB memory system has 10 levels, i.e., integrity verification can increase memory bandwidth demands by an order of magnitude.

A *Bonsai Merkle Tree* (BMT) helps reduce the size of the Merkle Tree data structure [258]. Every data block is associated with a counter or version number that is used when encrypting/decrypting the data block. To correctly read a block, the processor needs the data block, its counter, and its MAC. To carry out a replay attack, the attacker needs to somehow furnish an old version of the block, counter, and MAC. By constructing a Merkle Tree over the counters, we can guarantee that the counter value is fresh, thus thwarting a replay attack. The advantage of doing this is that counters require fewer bits; we are able to therefore pack several counters into a leaf node of the Merkle Tree, thus reducing its depth. This gives us a Bonsai Merkle Tree, which differs from a Merkle Tree in the arity and contents of the last level of the tree. Figure 11.4 shows example organizations of a Merkle and Bonsai Merkle Tree.

If we use 8 bits per counter, the last level of the BMT has an arity of 64. This improves the depth of the integrity tree from 10 to 9 for our 64-GB example memory system. However, note that a counter increments every time the corresponding block is written, i.e., 8-bit counter values are recycled after every 256 writes. This would enable a replay attack. To eliminate this possibility, we would need much larger counters, which in turn would lower the arity of the BMT's last level. Rogers et al. [258] address this by maintaining a large shared global counter value per memory block. For example, a 512-bit memory block can contain a 64-bit shared global counter and 64 7-bit local counters. We can create 64 71-bit counters by concatenating the global counter and each local counter. This gives us sufficiently large counter values and a sufficiently high arity. Every time a local counter rotates back to zero, the global counter is incremented and all other local counters are also reset. This local counter overflow is an expensive (albeit infrequent) operation as it requires re-encrypting all 64 data blocks that share the global counter.

Another performance optimization is to cache part of the BMT in the processor's cache hierarchy. When fetching BMT ancestors of a memory block, we can stop our upward traversal as soon as we encounter a cache hit (since the attacker cannot modify that cached node). If we assume that the top 5 levels of the BMT (requiring about 2 MB of storage) can be cached on average, it roughly halves the memory bandwidth requirements of integrity verification.

Intel SGX

Intel's SGX architecture incorporates several interesting security features [259, 260]. Here, we will briefly discuss how SGX handles security within its memory system, with integrity verification serving as the centerpiece. For more details on SGX, please refer to the authoritative treatise by Costan and Devadas [259].

A sensitive application operates within a "container" called an *enclave*, which is a region of memory with CIA guarantees—*Confidentiality*, *Integrity*, and *Authentication*. The hardware encrypts data written to the enclave and ensures that only the owner application has permissions to access this data, which to a large extent addresses Confidentiality and Authentication. But a malicious OS or an attacker with access to hardware can control the contents of memory—the Integrity guarantee ensures that the sensitive application can detect any such tampering. Note that the CIA guarantees only apply to sensitive pages in an enclave; non-sensitive pages can be accessed similar to a baseline non-secure memory.

The sensitive portion of main memory is partitioned into two parts—an *Enclave Page Cache* (EPC) and a non-EPC region. The EPC has a total capacity of only 128 MB, of which 32 MB is used for various metadata. So an application can only place 24 K, 4 KB sensitive pages in its EPC—the remaining sensitive pages are in the non-EPC region. Only the data in EPC

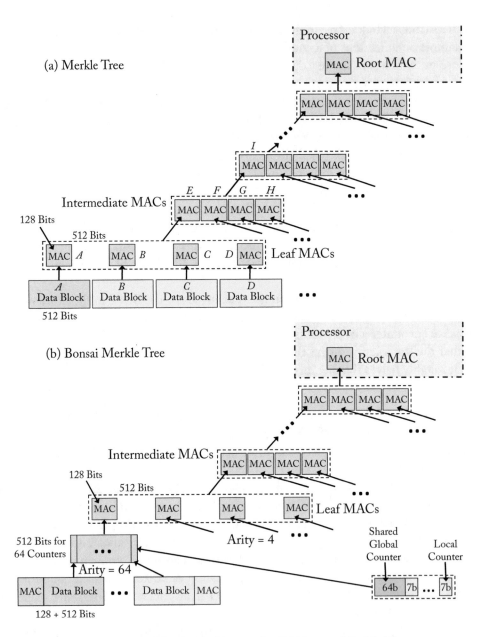

Figure 11.4: A comparison of a (a) Merkle Tree and a (b) Bonsai Merkle Tree.

can be directly accessed; if a requested sensitive block is in the non-EPC region, that page has to first be moved into EPC before the block can be accessed. This can lead to non-trivial paging overheads when an application has a sensitive working set size of more than 96 MB [261, 262].

The data in EPC is protected with an integrity tree organization shown in Figure 11.5. Every node in the tree is composed of eight 56-bit counters and a 64-bit hash. The hash is a function of the eight counters in that node and one 56-bit counter in the parent. The overall organization is thus very similar to the BMT, i.e., a tree with arity eight. Both use a hash to establish linkage between parent and child, but while BMT places the hash in the parent, SGX places the hash in the child. We will shortly compare the performance of BMT and SGX integrity structures.

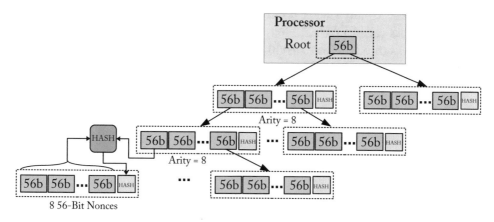

Figure 11.5: The integrity tree used by Intel SGX.

The cost of integrity verification for an EPC access is non-trivial. This cost is kept in check with a relatively small EPC region that in turn leads to an integrity tree with relatively low depth. The cost of a non-EPC access is even higher (about 40 K cycles [262]), requiring OS overheads, integrity overheads, and page movement between EPC and non-EPC regions. The non-EPC region is also protected with an integrity tree called the *eviction tree*. This tree is managed/accessed at the granularity of a page, i.e., counters/MACs are maintained per page to reduce storage overheads. Thus, all sensitive data blocks in an application's enclave are protected with integrity trees, either in the EPC region or in the non-EPC region.

It is worth noting that integrity imposes a memory storage penalty as well. The data structures required for integrity verification (MACs, counters, tree of hashes) can impose a storage penalty of about 15–25% for Merkle and Bonsai Merkle Trees [263]. But the storage penalty in SGX is under 1% because the EPC integrity tree only handles 96 MB of data, while the non-EPC integrity tree is managed at page granularity.

Thus, SGX has created an embodiment of integrity verification that brings it out of the shadows of academic research. Instead of the order of magnitude slowdown associated with integrity, SGX offers a palatable integrity tree for a small EPC region. Unfortunately, that also ends up being its Achilles Heel. The integrity overheads increase dramatically when the sensitive working set is larger than the EPC. Moving forward, we will either need applications that can define a small sensitive working set, or design architectures to improve SGX's scalability—we discuss such architectures next.

As a closing thought, note that SGX does not defend against a number of side channels, including the timing and address side channels discussed in Sections 11.1 and 11.2. A malicious OS on SGX can also easily observe the pattern of pages being touched by an application [246–248].

VAULT

To address the scalability challenges in SGX, Taassori et al. [263] introduced a collection of integrity techniques called VAULT. First, the EPC and non-EPC regions are unified, thus eliminating paging overheads. However, this increases the overheads of the integrity tree. To keep that overhead in check, VAULT uses a new integrity tree structure shown in Figure 11.6.

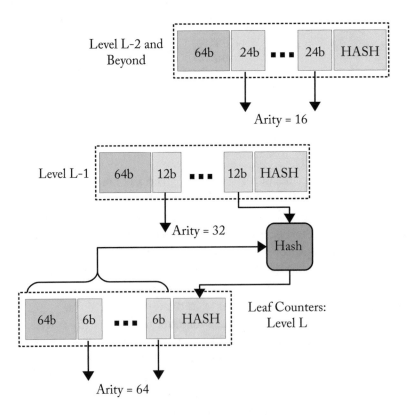

Figure 11.6: The integrity tree in VAULT.

Similar to SGX, it uses a hash to establish parent-child linkage. But instead of the 56-bit counters used by SGX, it uses a global 64-bit counter and several few-bit local counters per node. This helps increase the arity of the tree. However, as discussed earlier, significant overheads are incurred every time a local counter overflows. There is thus a trade-off between arity and overflow. Each level of the tree has a different optimal trade-off point—high arity is fine in lower levels, but higher levels that see more frequent counter updates need larger counters. This leads to a variable-arity tree organization.

Finally, to reduce storage overheads, Taassori et al. compress the block and insert the MAC into the space thus created, i.e., a single 64-byte block accommodates a compressed version of the block and its MAC. This approach is less effective when blocks aren't compressible. Multiple such blocks can share a MAC to lower their storage overheads, while incurring a memory bandwidth penalty.

Thus, VAULT is able to grow the sensitive working set size, while keeping the integrity tree depth and the storage overheads in check.

Morph

Saileshwar et al. [264] introduce new techniques to pack more counters per cache line and shrink the size of the integrity tree. They assume a baseline where counters in a line are decomposed into a shared global counter and few-bit local counters. They observe that pages largely exhibit two types of behaviors: either few lines in the page see frequent write activity, or nearly all lines in the page are accessed uniformly. In case of the former, a bit vector is used to track the few non-zero local counters and these few counters receive more than their fair share of bits. This helps reduce local counter overflows. In case of the latter, the counter values are stored in terms of an offset to a base. If the counters are updated uniformly, the base is periodically updated, thus keeping the offset small. Both of these techniques make it possible to increase the arity of the integrity tree and reduce the bandwidth overhead of integrity verification.

Synergy

Once the integrity tree is made more efficient with the techniques just described, the bottleneck in integrity verification shifts. The overhead of fetching a MAC becomes more significant, especially if the MAC working set size is large and MACs are frequently not found in the LLC. Saileshwar et al. [265] make the neat observation that most systems that need security also need error correction. While ECC is part of the memory rank and does not require a separate bus transaction, the MAC fetch does require a separate bus transaction. The authors therefore swap the MAC and the ECC; the MAC is fetched along with the block on a (say) 72-wide memory channel. If the block does have a hard or soft error, with a high probability, the MAC check will not succeed. This serves as an error detection. To correct the error, the separately stored ECC bits are retrieved. Thus, in the common case, integrity verification for reads does not impose any bandwidth overhead beyond that of an ECC-capable memory channel. However, for writes, in addition to updating the data block and its MAC, the separately stored ECC must also be updated. Therefore, writes would impose the same bandwidth overhead as the baseline technique for integrity verification. The separately stored ECC can use a parity function that can support chipkill recovery, i.e., correction capability even when an entire chip fails.

11.4   IMPACT OF SMART MEMORIES

The ORAM and integrity tree approaches described in the last two sections have very high memory bandwidth and memory capacity overheads that have impeded commercial adoption. These solutions are based on the premise that memory is dumb and untrusted. This has been a reasonable premise for decades when memory vendors resisted logic additions to memory devices. But with the emerging market for smart memories and example products like Micron's HMC [33], that premise is being disrupted to some extent.

Let's assume that we have a memory device like the HMC, where (i) the logic layer can perform the necessary cryptographic functions, (ii) the package and its internal TSVs/chips cannot be easily examined or tampered with, and (iii) the memory device can be added to the trusted computing base. How easy is it to provide integrity and leakage-free guarantees?

We are thus assuming that the memory manufacturer is not malicious. Standard practices and public key infrastructures can be used to authenticate a memory device and establish a secure connection between the processor and memory device. Once this is done, there are two possible vulnerabilities that we have to defend against: (i) an attacker can observe the processor/memory channel and deduce access patterns; and (ii) an attacker can modify the hardware to (for example) set up a person-in-the-middle attack and possibly disrupt execution.

Two concurrent works, ObfusMem [266] and InvisiMem [267] constructed defenses for these attacks, thus guaranteeing integrity and zero leakage of access patterns. We will first describe the ObfusMem approach and then point out the differences in InvisiMem.

ObfusMem [266] first encrypts data payloads with the processor's private key and a version number. It then encrypts the data and address/command with counter-mode encryption. The receiving memory device decrypts the data and address/command; it then performs the necessary operation, either storing the encrypted data into the specified address, or reading encrypted data from that address and returning it to the processor. Since we are using counter-mode encryption for the processor/memory channel, an attacker cannot detect either spatial or temporal locality in the stream of encrypted addresses. Since the data payload is double-encrypted, the attacker can also not detect multiple transmissions of the same or similar data blocks. There is thus no leakage of access patterns to an external observer. To prevent leakage of the read/write ratio, ObfusMem also enforces that every read transaction is followed by a write transaction (essentially writing the block evicted by a cache fill). To ensure that all memory devices are accessed uniformly, dummy transactions are inserted so that all devices are busy when at least one device is accessed.

The encryption and decryption do not add much delay to the memory access. In counter-mode encryption, the one-time pad can be pre-computed and the only delay on the critical path is the XOR of the payload with the one-time pad.

To provide authentication guarantees, since the memory device cannot be tampered with, we must only ensure that any changes on the processor-memory channel are detected. Every message payload therefore includes a MAC before the counter-mode encryption is performed. Since every transmission uses a new counter value, replays are not possible and any data modification is detected with very high probability.

InvisiMem [267] includes many of the same features as above. To further close timing channels, InvisiMem proposes that every memory device be accessed at a constant rate. Aga et al. [267] observe that this is not penal because the SerDes channels used in HMC-like devices must anyway send periodic null packets for synchronization. InvisiMem also includes a hash in the memory device in case data is maliciously modified with a row-hammer attack [268]—adding an ECC per block would provide similar functionality.

The ObfusMem and InvisiMem solutions are effective in handling the primary memory attacks at very low performance and storage overheads. Are they perfect solutions that render integrity trees and ORAM obsolete? The only catch is that 3D-stacked or active-memory devices are expected to be relatively expensive. Each package offers limited capacity at high cost and such memories have not yet been used to construct the entire memory system in a high-capacity server. Instead, they have been used primarily as a large DRAM cache (capacities up to 16 GB), with an additional hundreds of giga-bytes of memory supported with DDR-based DIMMs. Access to these DDR DIMMs will continue to need ORAM and integrity verification support for high security/privacy.

11.5   OTHER MEMORY SECURITY ISSUES

In this section, we examine a few other memory system vulnerabilities.

As NVMs become popular, data extraction from stolen NVDIMMs is a very credible threat. Encryption of NVDIMM data therefore becomes even more critical. Encrypted blocks have high entropy, i.e., on average, half the bits are flipped on any block write, which greatly impacts the energy and endurance of most NVM devices. To alleviate this effect, Young et al. introduce DEUCE, which tracks dirty bits at word granularity and only writes to modified words on every update [269]. Baseline encryption uses the block's address, version number, and the private key to generate a one time pad; the block is XOR-ed with the one time pad to generate an encrypted version of the block. DEUCE maintains two counters, a leading counter and a trailing counter. At the start of an epoch, a block is encrypted with the baseline approach using the trailing counter. When a block is modified, only the dirty words are modified using a one time pad generated by the leading counter. Thus, with a small increase in metadata, the number of cell modifications are reduced by 2×. Updates to the metadata is also a significant overhead in encrypted NVMs. While the standard solution is to cache metadata on the processor, this leads to the problem of crash inconsistency, where upon a crash, the NVDIMM may contain an updated block, but not its updated metadata. Recent work shows how to support low-cost crash consistency by exploiting application semantics [270] or ECC [271]. Another approach to reducing writes in encrypted NVMs is deduplication [272].

Row hammer is a memory security phenomenon that has received a lot of attention from popular media. This is a phenomenon that was observed by researchers at Intel in 2012 [273], and independently studied by other teams [268]. When a given row in DRAM is accessed repeatedly, the contents of nearby cells are affected. This is caused by voltage fluctuations in the DRAM wordline that accelerate leakage in nearby cells. Therefore, to disrupt a value, a single row must be accessed thousands of times while not accessing neighboring rows. The "bug" is especially problematic because a seemingly benign user thread in existing systems can disrupt the data of other threads without violating kernel privileges. Researchers have also shown more diabolical exploitations of this bug where a user thread can escalate its privilege levels [257].

Note that a row hammer attack violates data integrity. But it's an attempt to randomly disrupt bits, i.e., an attacker cannot precisely construct an effective replay attack. Such an attack can therefore be easily detected by having an ECC or MAC per block. This doesn't mean that a row hammer attack is harmless—recovering the corrupted data is much harder because more bits in a word can be corrupted than the correction capabilities of the ECC.

To prevent a row hammer attack, either repeated access to a row must be prevented, or neighboring rows must be accessed periodically to replenish charge in their cells. Kim et al. [268] suggest a number of mitigation approaches for row hammer. One of the techniques is especially effective and does not require any metadata or book-keeping overheads [268, 274]. When a row is activated, with a small probability, the memory controller also activates a neighboring row. Thus, the probability of a row not being replenished while its neighbor is repeatedly accessed is exponentially tiny. The performance overhead is small—it's a function of the probability of a neighbor activation. The memory controller also does not require any metadata tables to track activity per row. Alternatively, a set of counters can track rows that may be subjected to hammering [274]. Seyedzadeh et al. [275] use a tree data structure to reduce the number of counters required to track the most frequently accessed rows. The work of Aweke et al. [276] uses software monitoring of hardware program counters on existing systems to identify such aggressor rows.

While row hammer mitigation is simple and will likely be included in future memory controllers, it created waves primarily because of the large number of in-the-field systems that did not have mitigation techniques and that were vulnerable to attacks.

11.6   DISCUSSION

Intel SGX and other trusted execution environments provide a clue about what industry is willing to pay for high security. In the best case, SGX provides a slowdown of nearly 2× because of the overhead imposed by integrity verification. Therefore, techniques that eliminate leakage must impose overheads that are in a similar ball-park to be commercially viable. We've discussed timing-channel-free memory controllers that impose a 2× slowdown; so they may be palatable for industry. In any case, a 2× overhead for integrity and for zero timing channels is in itself a very high overhead and deserving of more research. Integrity appears to be the easier problem, thanks to the many recent works in this area. A timing-channel-free and overhead-free scheduler, on the other hand, seems elusive.

Efficient ORAM also appears elusive. This is because it is not amenable to most system optimizations that take advantage of locality; ORAMs either do not exhibit much locality or must exhibit worst-case behavior to prevent leakage. More innovation is clearly required before ORAM can be commercially attractive. Smart memory devices can help, but will not be enough as long as commodity DDR devices continue to provide a significant cost-per-bit advantage. Efficient memory security is therefore an important challenge for the next decade.

CHAPTER 12

Closing Thoughts

We are in the middle of a transition from a memory industry singularly focused on density to a memory industry up against the scaling wall and looking for value additions.

Some of the path ahead is very clear. There is a large investment in NVMs that will continue to scale for a few more generations. These NVMs will provide significant capacity enhancements, but they will likely not make DRAM obsolete. The problems faced by traditional memory systems will be amplified with NVMs. NVMs will also introduce new problems, e.g., very high latencies for writes, endurance, drift, etc.

Memory hierarchies will also get deeper. In addition to DDR-based DRAM DIMMs, the hierarchy will incorporate longer-latency NVMs and lower-latency 3D-stacked HBM or HMC-like devices. Much work remains in defining optimal policies for these deep hierarchies. Similarly, more work is required to reduce the cost of data movement. This will require innovative strategies for on-chip, on-DIMM, and inter-module interconnects and data codes.

Projections show that memory errors will continue to get worse. While we have a handle on efficiently tolerating a few errors, it is unclear just how error-prone future memories will be and how strong the codes need to be. Similarly, refresh penalties are also rising; while refresh latency can be hidden thanks to recent breakthroughs, refresh energy will continue to be a noticeable overhead.

In my opinion, we are unlikely to see big performance jumps with better memory scheduling algorithms. It is also unlikely that academics will stumble upon new DRAM microarchitectures that will dramatically improve latency, energy, and bandwidth. Perhaps technology shifts will inject new life into these areas.

I view three major topics as being especially fertile research areas. Memory compression has a lot to offer and commercial systems haven't yet tapped into most of these benefits. In the future, I expect to see solutions that unify the handling of compression, reliability, and security. Which brings me to the second major fertile topic: memory security. Thanks to Spectre and Meltdown, side-channel defenses will no longer be ignored. The security cat-and-mouse game will continue for the foreseeable future, giving rise to new attacks, and requiring new solutions. Even for known solutions (ORAM and timing-channel-free memory controllers), the room for improvement is very significant. Finally, near data processing remains a grand challenge. Future computing systems will inevitably have compute and auxiliary operations scattered all across a node. As more prototype systems emerge, the programming framework and killer applications will take shape. This will expose new bottlenecks and problems. We are also likely to see many more in-memory operators and many more accelerators that exploit NDP principles in their design. I expect that the community will take big strides in realizing a feature-rich memory system in the coming decade.

Bibliography

[1] T. Pawlowski. Hybrid memory cube (HMC). In *HotChips*, 2011. DOI: 10.1109/hotchips.2011.7477494 18, 19, 70

[2] R. Balasubramonian, A. B. Kahng, N. Muralimanohar, A. Shafiee, and V. Srinivas. CACTI 7: New tools for interconnect exploration in innovative off-chip memories. *ACM TACO*, 14(2), 2017. DOI: 10.1145/3085572 13, 16, 41, 44, 45, 48

[3] N. Chatterjee, R. Balasubramonian, M. Shevgoor, S. Pugsley, A. Udipi, A. Shafiee, K. Sudan, M. Awasthi, and Z. Chishti. USIMM: the Utah simulated memory module. *Technical Report*, University of Utah, 2012. 11, 12

[4] B. Jacob, S. W. Ng, and D. T. Wang. *Memory Systems—Cache, DRAM, Disk*. Elsevier, 2008. xix, 1, 11, 12

[5] GDDR6: The Next-Generation Graphics DRAM. https://www.micron.com/-/media/client/global/documents/products/technical-note/dram/tned03_gddr6.pdf, 2017. 15, 16

[6] A. Shilov. SK Hynix Details DDR5–6400, 2019. https://www.anandtech.com/show/13999/sk-hynix-details-its-ddr56400-dram-chip 5, 13

[7] L. Armasu. What We Know About DDR5 So Far, 2019. https://www.tomshardware.com/news/what-we-know-ddr5-ram,39079.html 13, 14, 16

[8] J. Mukundan, H. Hunter, K.-H. Kim, J. Stuecheli, and J. F. Martinez. Understanding and mitigating refresh overheads in high-density DDR-4 DRAM systems. In *Proc. of ISCA*, 2013. DOI: 10.1145/2508148.2485927 61, 62, 63, 66

[9] M. Qureshi, S. Gurumurthi, and B. Rajendran. *Phase Change Memory: From Devices to Systems*. Morgan & Claypool Synthesis Lectures on Computer Architecture, 2011. DOI: 10.2200/s00381ed1v01y201109cac018 1, 4

[10] Tilera. Tilera Tile64 Product Brief. http://www.tilera.com/sites/default/files/productbriefs/PB010_TILE64_Processor_A_v4.pdf 13

[11] M. J. Miller. Sandy Bridge-E: The Fastest Desktop Chip Ever (For Now). http://forwardthinking.pcmag.com/none/290775-sandy-bridge-e-the-fastest-desktop-chip-ever-for-now

[12] Intel. Products Formerly Skylake. https://ark.intel.com/content/www/us/en/ark/products/codename/37572/skylake.html 13

[13] ITRS. International Technology Roadmap for Semiconductors, 2009. 13

[14] Registered DIMMs. http://www.micron.com/products/dram-modules/rdimm 14

[15] Load-Reduced DIMMs. http://www.micron.com/products/dram-modules/lrdimm 14, 91

[16] Rambus. Get Ready for DDR5 DIMM Chipsets, 2019. https://www.rambus.com/blogs/get-ready-for-ddr5-dimm-chipsets/ 14

[17] J. Ahn, N. Jouppi, and R. S. Schreiber. Future scaling of processor-memory interfaces. In *Proc. of SC*, 2009. DOI: 10.1145/1654059.1654102 15, 37, 38, 48

[18] H. Zheng, J. Lin, Z. Zhang, E. Gorbatov, H. David, and Z. Zhu. Mini-rank: Adaptive DRAM architecture for improving memory power efficiency. In *Proc. of MICRO*, 2008. DOI: 10.1109/micro.2008.4771792 37, 38, 48

[19] A. N. Udipi, N. Muralimanohar, N. Chatterjee, R. Balasubramonian, A. Davis, and N. Jouppi. Rethinking DRAM design and organization for energy-constrained multi-cores. In *Proc. of ISCA*, 2010. DOI: 10.1145/1815961.1815983 15, 35, 37, 56

[20] D. Lee, M. O'Connor, and N. Chatterjee. Reducing data transfer energy by exploiting similarity within a data transaction. In *Proc. of HPCA*, 2018. DOI: 10.1109/hpca.2018.00014 15, 46

[21] P. Vogt. Fully buffered DIMM (FB-DIMM) server memory architecture: Capacity, performance, reliability, and longevity. Intel Developer Forum, 2004. 16

[22] J. Lin, H. Zheng, Z. Zhu, Z. Zhang, and H. David. DRAM-level prefetching for fully-buffered DIMM: Design, performance and power saving. In *Proc. of ISPASS*, 2007. DOI: 10.1109/ispass.2007.363740 16

[23] B. Ganesh, A. Jaleel, D. Wang, and B. Jacob. Fully-buffered DIMM memory architectures: Understanding mechanisms, overheads, and scaling. In *Proc. of HPCA*, 2007. DOI: 10.1109/hpca.2007.346190 16

[24] B. Mutnury, M. Cases, N. Pham, D. deAraujo, E. Matoglu, P. Patel, and B. Hermann. Analysis of fully buffered DIMM interface in high-speed server applications. In *Proc. of Electronic Components and Technology Conference*, 2006. DOI: 10.1109/ectc.2006.1645648 16

[25] Intel. Intel 7500/7510/7512 scalable memory buffer (Datasheet). *Technical Report*, 2011. 16, 18

[26] S. Pawlowski. Intelligent and expandable high-end intel server platform, codenamed nehalem-EX. *Intel Developer Forum*, `http://blogs.intel.com/technology/Nehal em-EX_Steve_Pawlowski_IDF.pdf`, 2009.

[27] D. Watts, D. Furniss, S. Haddow, J. Jervay, E. Kern, and C. Knight. IBM eX5 Portfolio Overview: IBM System x3850 X5, x3950 X5, x3690 X5, and BladeCenter HX5. `www. redbooks.ibm.com/abstracts/redp4650.html` 16

[28] S. Behling, R. Bell, P. Farrell, H. Holthoff, F. O'Connell, and W. Weir. The POWER4 Processor Introduction and Tuning Guide. `http://www.redbooks.ibm.com/redbook s/pdfs/sg247041.pdf` 18

[29] E. Cooper-Balis, P. Rosenfeld, and B. Jacob. Buffer on board memory systems. In *Proc. of ISCA*, 2012. DOI: 10.1109/isca.2012.6237034 18

[30] Samsung. Samsung to Release 3D Memory Modules with 50% Greater Density, 2010. `http://www.computerworld.com/s/article/9200278/Samsung_to_relea se_3D_memory_modules_with_50_greater_density` 18

[31] Tezzaron Semiconductor. 3D Stacked DRAM/Bi-STAR Overview, 2011. `http://ww w.tezzaron.com/memory/Overview_3D_DRAM.htm`

[32] Elpida Memory Inc. News Release: Elpida, PTI, and UMC Partner on 3D IC Integration Development for Advanced Technologies Including 28 nm. `http://www.elpida .com/en/news/2011/05--30.html`, 2011. 18

[33] J. Jeddeloh and B. Keeth. Hybrid memory cube–new DRAM architecture increases density and performance. In *Symposium on VLSI Technology*, 2012. DOI: 10.1109/vlsit.2012.6242474 18, 28, 43, 98

[34] T. Shimizu. Fujitsu HPC Roadmap Beyong Petascale Computing, 2013. `www.fujitsu.com/downloads/TC/sc13/fujitsu-hpc-roadmap-beyond-petascale-computing.pdf` 18

[35] JEDEC. High bandwidth memory (HBM) DRAM, 2013. JESD235. 19

[36] S. Rixner, W. Dally, U. Kapasi, P. Mattson, and J. Owens. Memory access scheduling. In *Proc. of ISCA*, 2000. DOI: 10.1109/isca.2000.854384 22

[37] K. J. Nesbit, N. Aggarwal, J. Laudon, and J. E. Smith. Fair queuing memory systems. In *Proc. of MICRO*, 2006. DOI: 10.1109/micro.2006.24 22

[38] O. Mutlu and T. Moscibroda. Stall-time fair memory access scheduling for chip multiprocessors. In *Proc. of MICRO*, 2007. DOI: 10.1109/micro.2007.21 22

[39] O. Mutlu and T. Moscibroda. Parallelism-aware batch scheduling—enhancing both performance and fairness of shared DRAM systems. In *Proc. of ISCA*, 2008. DOI: 10.1109/ISCA.2008.7 23

[40] Y. Kim, M. Papamichael, O. Mutlu, and M. Harchol-Balter. Thread cluster memory scheduling: Exploiting differences in memory access behavior. In *Proc. of MICRO*, 2010. DOI: 10.1109/micro.2010.51 23

[41] E. Ipek, O. Mutlu, J. F. Martinez, and R. Caruana. Self optimizing memory controllers: A reinforcement learning approach. In *Proc. of ISCA*, 2008. DOI: 10.1109/isca.2008.21 23

[42] J. Mukundan and J. F. Martinez. MORSE: Multi-objective reconfigurable SElf-optimizing memory scheduler. In *Proc. of HPCA*, 2012. DOI: 10.1109/hpca.2012.6168945 23

[43] L. Subramanian, D. Lee, V. Seshadri, H. Rastogi, and O. Mutlu. BLISS: Balancing performance, fairness and complexity in memory access scheduling. *IEEE TPDS*, (27(10)), 2016. DOI: 10.1109/tpds.2016.2526003 24

[44] M. N. Bojnordi and E. Ipek. PARDIS: A programmable memory controller for the DDRx interfacing standards. In *Proc. of ISCA*, 2012. DOI: 10.1109/isca.2012.6237002 24, 27

[45] J. Stuecheli, D. Kaseridis, D. Daly, H. Hunter, and L. John. The virtual write queue: Coordinating DRAM and last-level cache policies. In *Proc. of ISCA*, 2010. DOI: 10.1145/1815961.1815972 24

[46] N. Chatterjee, N. Muralimanohar, R. Balasubramonian, A. Davis, and N. Jouppi. Staged reads: Mitigating the impact of DRAM writes on DRAM reads. In *Proc. of HPCA*, 2012. DOI: 10.1109/hpca.2012.6168943 24

[47] C. J. Lee, O. Mutlu, V. Narasiman, and Y. N. Patt. Prefetch-aware DRAM controllers. In *Proc. of MICRO*, 2008. DOI: 10.1109/micro.2008.4771791 25

[48] D. Kaseridis, J. Stuecheli, , and L. K. John. Minimalist open-page: A DRAM page-mode scheduling policy for the many-core era. In *Proc. of MICRO*, 2011. DOI: 10.1145/2155620.2155624 25, 27, 37

[49] Y. Kim, D. Han, O. Mutlu, and M. Harchol-Balter. ATLAS: A scalable and high-performance scheduling algorithm for multiple memory controllers. In *Proc. of HPCA*, 2010. DOI: 10.1109/hpca.2010.5416658 25

[50] M. Awasthi, D. Nellans, K. Sudan, R. Balasubramonian, and A. Davis. Handling the problems and opportunities posed by multiple on-chip memory controllers. In *Proc. of PACT*, 2010. DOI: 10.1145/1854273.1854314 25, 27

[51] K. Sudan, N. Chatterjee, D. Nellans, M. Awasthi, R. Balasubramanian, and A. Davis. Micro-pages: Increasing DRAM efficiency with locality-aware data placement. In *Proc. of ASPLOS-XV*, 2010. DOI: 10.1145/1735970.1736045 27, 37

[52] V. Seshadri, T. Mullins, A. Boroumand, O. Mutlu, P. Gibbons, M. Kozuch, and T. Mowry. Gather-scatter DRAM: In-DRAM address translation to improve the spatial locality of non-unit strided accesses. In *Proc. of MICRO*, 2015. DOI: 10.1145/2830772.2830820 27

[53] J. Carter, W. Hsieh, L. Stroller, M. Swanson, L. Zhang, E. L. Brunvand, A. Davis, C.-C. Kuo, R. Kuramkote, M. A. Parker, L. Schaelicke, and T. Tateyama. Impulse: Building a smarter memory controller. In *Proc. of HPCA*, 1999. DOI: 10.1109/hpca.1999.744334 27

[54] Z. Zhang, Z. Zhu, and X. Zhand. A permutation-based page interleaving scheme to reduce row-buffer conflicts and exploit data locality. In *Proc. of MICRO*, 2000. DOI: 10.1109/micro.2000.898056 27

[55] B. Lin, M. B. Healy, R. Miftakhutdinov, P. G. Emma, and Y. Patt. Duplicon cache: Mitigating off-chip memory bank and bank group conflicts via data duplication. In *Proc. of MICRO*, 2018. DOI: 10.1109/micro.2018.00031 27

[56] J. Kotra, H. Zhang, A. Alameldeen, C. Wilkerson, and M. Kandemir. Chameleon: A dynamically reconfigurable heterogeneous memory system. In *Proc. of MICRO*, 2018. DOI: 10.1109/micro.2018.00050 27

[57] N. Chatterjee, M. O'Connor, D. Lee, D. Johnson, S. W. Keckler, M. Rhu, and W. J. Dally. Architecting an energy-efficient DRAM system for GPUs. In *Proc. of HPCA*, 2017. DOI: 10.1109/hpca.2017.58 28, 35, 37, 38

[58] R. Tremaine, P. Franaszek, J. Robinson, C. Schulz, T. Smith, M. Wazlowski, and P. Bland. IBM memory expansion technology (MXT). *IBM Journal of Research and Development*, 45(2), 2001. DOI: 10.1147/rd.452.0271 28

[59] B. Abali, H. Franke, S. Xiaowei, D. Poff, and T. Smith. Performance of hardware compressed main memory. In *Proc. of HPCA*, 2001. DOI: 10.1109/hpca.2001.903253 28

[60] P. A. Franaszek, J. Robinson, and J. Thomas. Parallel compression with cooperative dictionary construction. In *Proc. of the Data Compression Conference*, 1996. DOI: 10.1109/dcc.1996.488325 28

[61] M. Ekman and P. Stenstrom. A robust main-memory compression scheme. In *Proc. of ISCA*, 2005. DOI: 10.1109/isca.2005.6 29

[62] A. R. Alameldeen and D. A. Wood. Frequent pattern compression: A significance-based compression scheme for L2 caches. *Technical Report*, University of Wisconsin–Madison, 2004. 29

[63] J. Yang, Y. Zhang, and R. Gupta. Frequent value compression in data caches. In *Proc. of MICRO-33*, pages 258–265, December 2000. DOI: 10.1109/micro.2000.898076 29

[64] G. Pekhimenko, V. Seshadri, Y. Kim, H. Xin, O. Mutlu, M. A. Kozuch, P. B. Gibbons, and T. C. Mowry. Linearly compressed pages: A low-complexity, low-latency main memory compression framework. In *Proc. of MICRO*, 2013. DOI: 10.1145/2540708.2540724 29, 30

[65] G. Pekhimenko, V. Seshadri, O. Mutlu, P. B. Gibbons, M. A. Kozuch, and T. C. Mowry. Base-delta-immediate compression: Practical data compression for on-chip caches. In *Proc. of PACT*, 2012. DOI: 10.1145/2370816.2370870 29

[66] E. Choukse, M. Erez, and A. Alameldeen. Compresso: Pragmatic main memory compression. In *Proc. of MICRO*, 2018. DOI: 10.1109/micro.2018.00051 30

[67] A. Shafiee, M. Taassori, R. Balasubramonian, and A. Davis. MemZip: Exploiting unconventional benefits from memory compression. In *Proc. of HPCA*, 2014. DOI: 10.1109/hpca.2014.6835972 30, 31, 48

[68] V. Sathish, M. j. Schulte, and N. S. Kim. Lossless and lossy memory I/O link compression for improving performance of GPGPU workloads. In *Proc. of PACT*, 2012. DOI: 10.1145/2370816.2370864 31

[69] A. Deb, A. Shafiee, R. Balasubramonian, P. Faraboschi, N. Muralimanohar, and R. Schreiber. Enabling technologies for memory compression: Metadata, mapping, and prediction. In *Proc. of ICCD*, 2016. DOI: 10.1109/iccd.2016.7753256 30, 31

[70] D. Palframan, N. S. Kim, and M. Lipasti. COP: To compress and protect main memory. In *Proc. of ISCA*, 2015. DOI: 10.1145/2749469.2750377 31, 56

[71] S. Hong, P. Nair, B. Abali, A. Buyuktosunoglu, K-H. Kim, and M. B. Healy. Attache: Towards ideal memory compression by mitigating metadata bandwidth overheads. In *Proc. of MICRO*, 2018. DOI: 10.1109/micro.2018.00034 32

[72] V. Young, S. Kariyappa, and M. K. Qureshi. Enabling transparent memory-compression for commodity memory systems. In *Proc. of HPCA*, 2019. DOI: 10.1109/hpca.2019.00010 32

[73] Qualcomm. Qualcomm Centriq 2400 Processor, 2017. `https://www.qualcomm.com/m` `edia/documents/files/qualcomm-centriq-2400-processor.pdf` 32, 33

[74] D. Hepkin. *Active Memory Expansion: Overview and Usage Guide*, 2010. 32

[75] B. Blaner, B. Abali, B. M. Bass, S. Chari, R. Kalla, et al. IBM POWER7+ processor on-chip accelerators for cryptography and active memory expansion. *IBM Journal of Research and Development*, 57(6), 2013. DOI: 10.1147/jrd.2013.2280090 33

[76] E. Cooper-Balis and B. Jacob. Fine-grained activation for power reduction in DRAM. *IEEE Micro*, May/June 2010. DOI: 10.1109/mm.2010.43 35, 37

[77] T. Zhang, K. Chen, C. Xu, G. Sun, T. Wang, and Y. Xie. Half-DRAM: A high-bandwidth and low-power DRAM architecture from the rethinking of fine-grained activation. In *Proc. of ISCA*, 2014. DOI: 10.1145/2678373.2665724 35, 38

[78] H. Ha, A. Pedram, S. Richardson, S. Kvatinsky, and M. Horowitz. Improving energy efficiency of DRAM by exploiting half page row access. In *Proc. of MICRO*, 2016. DOI: 10.1109/micro.2016.7783730 36, 38

[79] M. O'Connor, N. Chatterjee, D. Lee, J. Wilson, A. Agrawal, S. Keckler, and W. Dally. Fine-grained DRAM: Energy-efficient DRAM for extreme bandwidth systems. In *Proc. of MICRO*, 2017. DOI: 10.1145/3123939.3124545 35, 37, 38, 42

[80] T. Vogelsang. Understanding the energy consumption of dynamic random access memories. In *Proc. of MICRO*, 2010. DOI: 10.1109/micro.2010.42 35

[81] Y. Lee, H. Kim, S. Hong, and S. Kim. Partial row activation for low-power DRAM system. In *Proc. of HPCA*, 2017. DOI: 10.1109/hpca.2017.35 39

[82] Y. Kim, V. Seshadri, D. Lee, J. Liu, and O. Mutlu. A case for exploiting subarray-level parallelism (SALP) in DRAM. In *Proc. of ISCA*, 2012. DOI: 10.1109/isca.2012.6237032 39, 61

[83] D. Lee, Y. Kim, V. Seshadri, J. Liu, L. Subramanian, and O. Mutlu. Tiered-latency DRAM: A low latency and low cost DRAM architecture. In *Proc. of HPCA-19*, 2013. DOI: 10.1109/hpca.2013.6522354 39

[84] Y. Son, S. O, Y. Ro, J. Lee, and J. Ahn. Reducing memory access latency with asymmetric DRAM bank organizations. In *Proc. of ISCA*, 2013. DOI: 10.1145/2508148.2485955 40

[85] M. Shevgoor, J-S. Kim, N. Chatterjee, R. Balasubramonian, A. Davis, and A. Udipi. Quantifying the relationship between the power delivery network and architectural policies in a 3D-stacked memory device. In *Proc. of MICRO*, 2013. DOI: 10.1145/2540708.2540726 40, 63

[86] V. Seshadri, Y. Kim, C. Fallin, D. Lee, R. Ausavarungnirun, G. Pekhimenko, Y. Luo, O. Mutlu, P. B. Gibbons, M. A. Kozuch, and T. C. Mowry. RowClone: Fast and energy-efficient in-DRAM bulk data copy and initialization. In *Proc. of MICRO*, 2013. DOI: 10.1145/2540708.2540725 40, 76

[87] S.-L. Lu, Y.-C. Lin, and C.-L. Yang. Improving DRAM latency with dynamic asymmetric subarray. In *Proc. of MICRO*, 2015. DOI: 10.1145/2830772.2830827 40

[88] K. Chang, P. Nair, D. Lee, S. Ghose, M. Qureshi, and O. Mutlu. Low-cost inter-linked subarrays (LISA): Enabling fast inter-subarray data movement in DRAM. In *Proc. of HPCA*, 2016. DOI: 10.1109/hpca.2016.7446095 40

[89] Y. Ro, H. Cho, E. Lee, D. Jung, Y. Son, J. Ahn, and J. Lee. SOUP-N-SALAD: Allocation-oblivious access latency reduction with asymmetric DRAM microarchitectures. In *Proc. of HPCA*, 2017. DOI: 10.1109/hpca.2017.31 40

[90] D. Lee, Y. Kim, G. Pekhimenko, S. Khan, V. Seshadri, K. Chang, and O. Mutlu. Adaptive-latency DRAM: Optimizing DRAM timing for the common case. In *Proc. of HPCA*, 2015. DOI: 10.1109/hpca.2015.7056057 40, 63

[91] K. Chandrasekar, S. Goossens, C. Weis, M. Koedam, B. Akesson, N. Wehn, and K. Goossens. Exploiting expendable process-margins in DRAMs for run-time performance optimization. In *Proc. of DATE*, 2014. DOI: 10.7873/date.2014.186 40, 63

[92] X. Zhang, Y. Zhang, B. R. Childers, and J. Yang. Exploiting DRAM restore time variations in deep sub-micron scaling. In *Proc. of DATE-15*, 2015. DOI: 10.7873/date.2015.0969 40

[93] M. Taassori, A. Shafiee, and R. Balasubramonian. Understanding and alleviating intra-die and intra-DIMM parameter variation in the memory system. In *Proc. of ICCD*, 2016. DOI: 10.1109/iccd.2016.7753283 40, 63

[94] S. Keckler. Life after dennard and how I learned to love the picojoule. Keynote at MICRO, 2011. 41

[95] X. Jian, P. Hanumolu, and R. Kumar. Understanding and optimizing power consumption in memory networks. In *Proc. of HPCA*, 2017. DOI: 10.1109/hpca.2017.60 41, 42, 46

[96] T. Schmitz. The rise of serial memory and the future of DDR, 2015. https://www.xilinx.com/support/documentation/white_papers/wp456-DDR-serial-mem.pdf 42

[97] K. T. Malladi, F. A. Nothaft, K. Periyathambi, B. C. Lee, C. Kozyrakis, and M. Horowitz. Towards energy-proportional datacenter memory with mobile DRAM. In *Proc. of ISCA*, 2012. DOI: 10.1145/2366231.2337164 42

[98] H. Wang, C.-J. Park, G.-S. Byun, J. H. Ahn, and N. S. Kim. Alloy: Parallel-serial memory channel architecture for single-chip heterogeneous processor systems. In *Proc. of HPCA*, 2015. DOI: 10.1109/hpca.2015.7056041 42, 43, 45

[99] G. Sandhu. DRAM scaling and bandwidth challenges. In *NSF Workshop on Emerging Technologies for Interconnects (WETI)*, 2012. 43

[100] J. T. Pawlowski. Hybrid memory cube (HMC). In *Hotchips*, 2011. DOI: 10.1109/hotchips.2011.7477494 43

[101] A. N. Udipi, N. Muralimanohar, R. Balasubramonian, A. Davis, and N. Jouppi. Combining memory and a controller with photonics through 3D-stacking to enable scalable and energy-efficient systems. In *Proc. of ISCA*, 2011. DOI: 10.1145/2024723.2000115 43, 44

[102] S. Beamer et al. Re-architecting DRAM memory systems with monolithically integrated silicon photonics. In *Proc. of ISCA*, 2010. DOI: 10.1145/1815961.1815978 43

[103] H. Seol, W. Shin, J. Jang, J. Choi, J. Suh, and L-S. Kim. Energy efficient data encoding in DRAM channels exploiting data value similarity. In *Proc. of ISCA*, 2016. DOI: 10.1109/isca.2016.68 47

[104] T. Nguyen, A. Fuchs, and D. Wentzlaff. CABLE: A cache-based link encoder for bandwidth-starved manycores. In *Proc. of MICRO*, 2018. DOI: 10.1109/micro.2018.00033 47

[105] Y. Song and E. Ipek. More is less: Improving the energy efficiency of data movement via opportunistic use of sparse codes. In *Proc. of MICRO*, 2015. DOI: 10.1145/2830772.2830806 47

[106] J. Gaur, M. Chaudhuri, P. Ramachandran, and S. Subramoney. Near-optimal access partitioning for memory hierarchies with multiple heterogeneous bandwidth sources. In *Proc. of HPCA*, 2017. DOI: 10.1109/hpca.2017.46 47

[107] G. Kim, J. Kim, J. Ahn, and J. Kim. Memory-centric system interconnect design with hybrid memory cubes. In *Proc. of PACT*, 2013. DOI: 10.1109/pact.2013.6618812 48

[108] G. Kim, M. Lee, J. Jeong, and J. Kim. Multi-GPU system design with memory networks. In *Proc. of MICRO*, 2014. DOI: 10.1109/micro.2014.55 48

[109] J. Zhan, I. Akgun, J. Zhao, A. Davis, P. Faraboschi, Y. Wang, and Y. Xie. A unified memory network architecture for in-memory computing in commodity servers. In *Proc. of MICRO*, 2016. DOI: 10.1109/micro.2016.7783732 48

[110] M. Poremba, I. Akgun, J. Yin, O. Kayiran, Y. Xie, and G. Loh. There and back again: Optimizing the interconnect in networks of memory cubes. In *Proc. of ISCA*, 2017. DOI: 10.1145/3140659.3080251 48

[111] M. Ogleari, Y. Yu, C. Qian, E. Miller, and J. Zhao. String figure: A scalable and elastic memory network architecture. In *Proc. of HPCA*, 2019. DOI: 10.1109/hpca.2019.00016 48

[112] D. H. Yoon, J. Chang, N. Muralimanohar, and P. Ranganathan. BOOM: Enabling mobile memory based low-power server DIMMs. In *Proc. of ISCA*, 2012. DOI: 10.1109/isca.2012.6237003 48

[113] J. Stuecheli. POWER8 Processor, 2013. https://www.hotchips.org/wp-content/uploads/hc_archives/hc25/HC25.20-Processors1-epub/HC25.26.210-POWER-Studecheli-IBM.pdf 48

[114] K. Lim, J. Chang, T. Mudge, P. Ranganathan, S. K. Reinhardt, and T. F. Wenisch. Disaggregated memory for expansion and sharing in blade servers. In *Proc. of ISCA*, 2009. DOI: 10.1145/1555754.1555789 49

[115] N. Jerger and L. Peh. *On-Chip Networks*. Morgan & Claypool Synthesis Lectures on Computer Architecture, 2009. DOI: 10.2200/s00209ed1v01y200907cac008 49

[116] B. Schroeder, E. Pinheiro, and W. D. Weber. DRAM errors in the wild: A large-scale field study. In *Proc. of SIGMETRICS*, 2009. DOI: 10.1145/1555349.1555372 52, 53

[117] V. Sridharan, N. DeBardeleben, S. Blanchard, K. B. Ferreira, J. Stearley, J. Shalf, and S. Gurumurthi. Memory errors in modern systems: The good, the bad, and the ugly. In *Proc. of ASPLOS*, 2015. DOI: 10.1145/2694344.2694348 52, 58

[118] C. Slayman et al. Impact of error correction code and dynamic memory reconfiguration on high-reliability/low-cost server memory. In *Integrated Reliability Workshop Final Report*, 2006. DOI: 10.1109/irws.2006.305243

[119] X. Li, M. C. Huang, K. Shen, and L. Chu. A realistic evaluation of memory hardware errors and software system susceptibility. In *Proc. of USENIX*, 2010.

[120] M. Blaum et al. The reliability of single-error protected computer memories. *IEEE Transactions on Computers*, 1988. DOI: 10.1109/12.75143

[121] S. Li, K. Chen, M. Y. Hsieh, N. Muralimanohar, C. D. Kersey, D. Chad, J. B. Brockman, A. F. Rodrigues, and N. P. Jouppi. System implications of memory reliability in exascale computing. In *SC*, 2011. DOI: 10.1145/2063384.2063445

[122] V. Sridharan and D. Liberty. A study of DRAM failures in the field. In *Proc. of SC*, 2013. DOI: 10.1109/sc.2012.13

[123] A. Hwang, I. Stefanovici, and B. Schroeder. Cosmic rays don't strike twice: Understanding the nature of DRAM errors and the implications for system design. In *Proc. of ASPLOS*, 2012. DOI: 10.1145/2248487.2150989 52

[124] S. Li, K. Chen, M.-Yu Hsieh, N. Muralimanohar, C. D. Kersey, J. B. Brockman, A. F. Rodrigues, and N. P. Jouppi. System implications of memory reliability in exascale computing. In *Proc. of International Conference for High Performance Computing, Networking, Storage and Analysis*, pages 46:1–46:12, 2011. DOI: 10.1145/2063384.2063445 53

[125] R. Hamming. Error detecting and error correcting codes. *Bell System Technical Journal*, 1950. DOI: 10.1002/j.1538-7305.1950.tb00463.x 53

[126] M. Y. Hsiao. A class of optimal minimum odd-weight-column SEC-DED codes. *IBM Journal of Research and Development*, 14, 1970. DOI: 10.1147/rd.144.0395 53

[127] J. Kim, M. Sullivan, and M. Erez. Bamboo ECC: Strong, safe, and flexible codes for reliable computer memory. In *Proc. of HPCA-15*, February 2015. DOI: 10.1109/hpca.2015.7056025 53, 54, 55

[128] Advanced Micro Devices (AMD), Inc. Kernel Developer's Guide for AMD NPT Family 0Fh Processors. http://developer.amd.com/wordpress/media/2012/10/325591.pdf 53

[129] Sun Microsystems, Inc. T2 Core Microarchitecture Specification. http://www.oracle.com/technetwork/systems/opensparc/t2--06-opensparct2-core-microarch-1537749.html 53

[130] Advanced Micro Devices (AMD), Inc. *BIOS and Kernel Developer's Guide (BKDG) for AMD Family 15h Models 00h-0Fh Processors*, 2013. 53, 54, 55

[131] Hewlett–Packard. *HP Advanced Memory Error Detection Technology*, 2011. 53

[132] S. Cha, S. O, H. Shin, S. Hwang, K. Park, S. J. Jang, J. S. Choi, G. Y. Jin, Y. H. Son, H. Cho, J. H. Ahn, and N. S. Kim. Defect analysis and cost-effective resilience architecture for future DRAM devices. In *Proc. of HPCA*, 2017. DOI: 10.1109/hpca.2017.30 54, 56, 57, 64

[133] P. Nair, D-H. Kim, and M. Qureshi. ArchShield: Architectural framework for assisting DRAM scaling by tolerating high error rates. In *Proc. of ISCA*, 2013. DOI: 10.1145/2508148.2485929 54, 56, 58, 63

[134] T. Y. Oh et al. A 3.2 Gb/s/pin 8 Gb 1.0V LPDDR4 SDRAM with integrated ECC engine for sub-1V DRAM core operation. In *Proc. of ISSCC*, 2014. DOI: 10.1109/isscc.2014.6757500 54, 57

[135] DDR4 Networking Design Guide, 2014. https://www.micron.com/~/media/docum ents/products/technical-note/dram/tn_4003_ddr4_network_design_guide.p df 54

[136] D. Yoon and M. Erez. Virtualized and flexible ECC for main memory. In *Proc. of ASPLOS*, 2010. DOI: 10.1145/1735970.1736064 54

[137] A. N. Udipi, N. Muralimanohar, R. Balasubramonian, A. Davis, and N. Jouppi. LOT-ECC: Localized and tiered reliability mechanisms for commodity memory systems. In *Proc. of ISCA*, 2012. DOI: 10.1145/2366231.2337192 55

[138] X. Jian, H. Duwe, J. Sartori, V. Sridharan, and R. Kumar. Low-power, low-storage-overhead chipkill correct via multi-line error correction. In *Proc. of SC*, 2013. DOI: 10.1145/2503210.2503243 55

[139] Y. H. Son, S. Lee, S. O, S. Kwon, N. S. Kim, and J. H. Ahn. CiDRA: A cache-inspired DRAM resilience architecture. In *Proc. of HPCA*, 2015. DOI: 10.1109/hpca.2015.7056058 56

[140] D. Kim and M. Erez. Balancing reliability, cost, and performance tradeoffs with FreeFault. In *Proc. of HPCA*, 2015. DOI: 10.1109/hpca.2015.7056053 56

[141] D. Kim and M. Erez. RelaxFault memory repair. In *Proc. of ISCA*, 2016. DOI: 10.1109/isca.2016.62 56

[142] S. H. Kim et al. A low power and highly reliable 400 Mbps mobile DDR SDRAM with on-chip distributed ECC. In *Proc. of ASSCC*, 2007. DOI: 10.1109/asscc.2007.4425789 56

[143] T. Nagai et al. A 65 nm low-power embedded DRAM with extended data-retention sleep mode. In *Proc. of ISSCC*, 2006. DOI: 10.1109/isscc.2006.1696093 56

[144] P. Nair, V. Sridharan, and M. Qureshi. XED: Exposing on-die error detection information for strong memory reliability. In *Proc. of ISCA*, 2016. DOI: 10.1145/3007787.3001174 56

[145] S.-L. Gong, J. Kim, S. Lym, M. Sullivan, H. David, and M. Erez. DUO: Exposing on-chip redundancy to rank-level ECC for high reliability. In *Proc. of HPCA*, 2018. DOI: 10.1109/hpca.2018.00064 57

[146] E. Ipek, J. Condit, E. Nightingale, D. Burger, and T. Moscibroda. Dynamically replicated memory: Building reliable systems from nanoscale resistive memories. In *Proc. of ASPLOS*, 2010. DOI: 10.1145/1735970.1736023 57

[147] S. Schechter, G. Loh, K. Strauss, and D. Burger. Use ECP, not ECC, for hard failures in resistive memories. In *Proc. of ISCA*, 2010. DOI: 10.1145/1815961.1815980 57

[148] N. H. Seong, D. H. Woo, V. Srinivasan, J. Rivers, and H. S. Lee. SAFER: Stuck-at-fault error recovery for memories. In *Proc. of MICRO*, 2010. DOI: 10.1109/micro.2010.46 57

[149] D-H Yoon, N. Muralimanohar, J. Chang, P. Ranganathan, N. Jouppi, and M. Erez. FREE-p: Protecting non-volatile memory against both hard and soft errors. In *Proc. of HPCA*, 2011. DOI: 10.1109/hpca.2011.5749752 57

[150] M. Awasthi, M. Shevgoor, K. Sudan, B. Rajendran, R. Balasubramonian, and V. Srinivasan. Efficient scrub mechanisms for error-prone emerging memories. In *Proc. of HPCA*, 2012. DOI: 10.1109/hpca.2012.6168941 57

[151] D. Zhang, V. Sridharan, and X. Jian. Exploring and optimizing chipkill-correct for persistent memory based on high-density NVRAMs. In *Proc. of MICRO*, 2018. DOI: 10.1109/micro.2018.00063 57

[152] M. Gupta, V. Sridharan, D. Roberts, A. Prodromou, A. Venkat, D. Tullsen, and R. Gupta. Reliability-aware data placement for heterogeneous memory architecture. In *Proc. of HPCA*, 2018. DOI: 10.1109/hpca.2018.00056 57

[153] X. Jian, V. Sridharan, and R. Kumar. Parity helix: Efficient protection for single-dimensional faults in multi-dimensional memory systems. In *Proc. of HPCA*, 2016. DOI: 10.1109/hpca.2016.7446094 57

[154] M. Shevgoor, R. Balasubramonian, N. Chatterjee, and J-S. Kim. Addressing service interruptions in memory with thread-to-rank assignment. In *Proc. of ISPASS*, 2016. DOI: 10.1109/ispass.2016.7482071 62, 66

[155] P. Nair, C. Chou, and M. Qureshi. A case for refresh pausing in DRAM memory systems. In *Proc. of HPCA*, 2013. DOI: 10.1109/hpca.2013.6522355 62

[156] K. Kim and J. Lee. A new investigation of data retention time in truly nanoscaled DRAMs. *IEEE Electron Device Letters*, 2009. DOI: 10.1109/led.2009.2023248 63

[157] J. Liu, B. Jaiyen, Y. Kim, C. Wilkerson, and O. Mutlu. An experimental study of data retention behavior in modern DRAM devices: Implications for retention time profiling mechanisms. In *Proc. of ISCA*, 2013. DOI: 10.1145/2508148.2485928 63

[158] M. Patel, J. S. Kim, and O. Mutlu. The reach profiler (REAPER): Enabling the mitigation of DRAM retention failures via profiling at aggressive conditions. In *Proc. of ISCA*, 2017. DOI: 10.1145/3140659.3080242 63, 64

[159] J. Liu, B. Jaiyen, R. Veras, and O. Mutlu. RAIDR: Retention-aware intelligent DRAM refresh. In *Proc. of ISCA*, 2012. DOI: 10.1109/isca.2012.6237001 63

[160] R. Venkatesan, S. Herr, and E. Rotenberg. Retention-aware placement in DRAM (RAPID): Software methods for quasi-non-volatile DRAM. In *Proc. of HPCA*, 2006. DOI: 10.1109/hpca.2006.1598122 63, 66

[161] S. Khan, C. Wilkerson, Z. Wang, A. Alameldeen, D. Lee, and O. Mutlu. Detecting and mitigating data-dependent DRAM failures by exploiting current memory content. In *Proc. of MICRO*, 2017. DOI: 10.1145/3123939.3123945 64

[162] M. K. Qureshi, D.-H. Kim, S. Khan, P. Nair J., and O. Mutlu. AVATAR: A variable-retention-time (VRT) aware refresh for DRAM systems. In *Proc. of 45th Annual IEEE/IFIP International Conference on Dependable Systems and Networks*, 2015. DOI: 10.1109/dsn.2015.58 64

[163] S. Khan, D. Lee, Y. Kim, A. Alameldeen, C. Wilkerson, and O. Mutlu. The efficacy of error mitigation techniques for DRAM retention failures: A comparative experimental study. In *Proc. of SIGMETRICS*, 2014. DOI: 10.1145/2637364.2592000 64

[164] C.-H. Lin, D.-Y. Shen, Y.-J. Chen, C.-L. Yang, and M. Wang. SECRET: Selective error correction for refresh energy reduction in DRAMs. In *Proc. of ICCD*, 2012. DOI: 10.1109/iccd.2012.6378619 64

[165] I. Bhati, Z. Chishti, S.-L. Lu, and B. Jacob. Flexible auto-refresh: Enabling scalable and energy-efficient DRAM refresh reductions. In *Proc. of ISCA*, 2015. DOI: 10.1145/2872887.2750408 64

[166] J. Stuecheli, D. Kaseridis, H. Hunter, and L. John. Elastic refresh: Techniques to mitigate refresh penalties in high density memory. In *Proc. of MICRO*, 2010. DOI: 10.1109/micro.2010.22 65

[167] M. Ghosh and H.-H. S. Lee. Smart refresh: An enhanced memory controller design for reducing energy in conventional and 3D die-stacked DRAMs. In *Proc. of MICRO*, 2007. DOI: 10.1109/micro.2007.13 65

[168] T. Zhang, M. Poremba, C. Xu, G. Sun, and Y. Xie. CREAM: A concurrent-refresh-aware DRAM memory architecture. In *Proc. of HPCA*, 2014. DOI: 10.1109/hpca.2014.6835947 65

[169] K. Kai-Wei Chang, D. Lee, Z. Chishti, C. Wilkerson, A. Alameldeen, Y. Kim, and O. Mutlu. Improving DRAM performance by parallelizing refreshes with accesses. In *Proc. of HPCA*, 2014. DOI: 10.1109/hpca.2014.6835946 65

[170] K. Nguyen, K. Lyu, X. Meng, V. Sridharan, and X. Jian. Nonblocking memory refresh. In *Proc. of ISCA*, 2018. DOI: 10.1109/isca.2018.00055 65

[171] C. Isen and L. John. ESKIMO—energy savings using semantic knowledge of inconsequential memory occupancy for DRAM subsystem. In *Proc. of MICRO*, 2009. DOI: 10.1145/1669112.1669156 66

[172] S. Liu, K. Pattabiraman, T. Moscibroda, and B. Zorn. Flikker: Saving DRAM refresh-power through critical data partitioning. In *Proc. of ASPLOS*, 2011. DOI: 10.1145/1961296.1950391 66

[173] J. Kotra, N. Shahidi, Z. Chishti, and M. Kandemir. Hardware-software co-design to mitigate DRAM refresh overheads. In *Proc. of ASPLOS*, 2017. DOI: 10.1145/3093336.3037724 66

[174] W. Shin, J. Yang, J. Choi, and L-S. Kim. NUAT: A non-uniform access time memory controller. In *Proc. of HPCA-20*, 2014. DOI: 10.1109/hpca.2014.6835956 67

[175] H. Hassan, G. Pekhimenko, N. Vijaykumar, V. Seshadri, D. Lee, O. Ergin, and O. Mutlu. ChargeCache: Reducing DRAM latency by exploiting row access locality. In *Proc. of HPCA*, 2016. DOI: 10.1109/hpca.2016.7446096 67

[176] X. Zhang, Y. Zhang, B. R. Childers, and J. Yang. Restore truncation for performance improvement in future DRAM systems. In *Proc. of HPCA*, 2016. DOI: 10.1109/hpca.2016.7446093 67

[177] Y. Wang, A. Tavakkol, L. Orosa, S. Ghose, N. Ghiasi, M. Patel, J. Kim, H. Hassan, M. Sadrosadati, and O. Mutlu. Reducing DRAM latency via charge-level-aware look-ahead partial restoration. In *Proc. of MICRO*, 2018. DOI: 10.1109/micro.2018.00032 67

[178] M. Alian, S. W. Min, H. Asgharimoghaddam, A. Dhar, D. K. Wang, T. Roewer, A. McPadden, O. O'Halloran, D. Chen, J. Xiong, D. Kim, W. Hwu, and N. S. Kim. Application-transparent near-memory processing architecture with memory channel network. In *Proc. of MICRO*, 2018. DOI: 10.1109/micro.2018.00070 69, 77

[179] B. Sukhwani, T. Roewer, C. L. Haymes, K-H. Kim, A. J. McPadden, D. M. Dreps, D. Sanner, J. V. Lunteren, and S. Asaad. ConTutto: A novel FPGA-based prototyping platform enabling innovation in the memory subsystem of a server class processor. In *Proc. of MICRO*, 2017. DOI: 10.1145/3123939.3124535 69

[180] T. Dysart, P. Kogge, M. Deneroff, E. Bovell, P. Briggs, J. Brockman, K. Jacobsen, Y. Juan, S. Kuntz, R. Lethin, J. McMahon, C. Pawar, M. Perrigo, S. Rucker, J. Ruttenberg, M. Ruttenberg, and S. Stein. Highly scalable near memory processing with migrating threads on the EMU system architecture. In *Proc. of the 6th Workshop on Irregular Applications: Architectures and Algorithms*, 2016. DOI: 10.1109/ia3.2016.007 69

[181] The Tomorrow Show: Three New Technologies from Hewlett Packard Labs, 2016. `https://youtu.be/tABpRpBW6h0?t=18m38s` 69

[182] K. Wang, M. Stan, and K. Skadron. Association rule mining with the micron automata processor. In *Proc. of IPDPS*, 2015. DOI: 10.1109/ipdps.2015.101 69

[183] S. Pugsley, J. Jestes, H. Zhang, R. Balasubramonian, V. Srinivasan, A. Buyuktosunoglu, A. Davis, and F. Li. NDC: Analyzing the impact of 3D-stacked memory+logic devices on MapReduce workloads. In *Proc. of ISPASS*, 2014. DOI: 10.1109/ispass.2014.6844483 70

[184] D. Zhang, N. Jayasena, A. Lyashevsky, J. Greathouse, L. Xu, and M. Ignatowski. TOP-PIM: Throughput-oriented programmable processing in memory. In *Proc. of HPDC*, 2014. DOI: 10.1145/2600212.2600213 70

[185] R. Nair et al. Active memory cube: A processing-in-memory architecture for exascale systems. *IBM Journal of R&D*, 59(2/3), 2015. DOI: 10.1147/JRD.2015.2409732 71

[186] A. Farmahini-Farahani, J. H. Ahn, K. Morrow, and N. S. Kim. NDA: Near-DRAM acceleration architecture leveraging commodity DRAM devices and standard memory modules. In *Proc. of HPCA-21*, 2015. DOI: 10.1109/hpca.2015.7056040 71

[187] V. Govindaraju, C.-H. Ho, T. Nowatzki, J. Chhugani, N. Satish, K. Sankaralingam, and C. Kim. DySER: Unifying functionality and parallelism specialization for energy efficient computing. *IEEE Micro*, 33(5), 2012. DOI: 10.1109/mm.2012.51 71

[188] M. Gao, G. Ayers, and C. Kozyrakis. Practical near-data processing for in-memory analytics frameworks. In *Proc. of PACT*, 2015. DOI: 10.1109/pact.2015.22 71

[189] S. Pugsley, J. Jestes, R. Balasubramonian, V. Srinivasan, A. Buyuktosunoglu, A. Davis, and F. Li. Comparing implementations of near data computing with in-memory MapReduce workloads. In *IEEE Micro's Special Issue on Big Data*, 2014. DOI: 10.1109/mm.2014.54 72

[190] H. Asghari-Moghaddam, Y. H. Son, J. H. Ahn, and N. S. Kim. Chameleon: Versatile and practical near-DRAM acceleration architecture for large memory systems. In *Proc. of MICRO*, 2016. DOI: 10.1109/micro.2016.7783753 72

[191] A. Shafiee, A. Nag, N. Muralimanohar, R. Balasubramonian, J. Strachan, M. Hu, R. S. Williams, and V. Srikumar. ISAAC: A convolutional neural network accelerator with in-situ analog arithmetic in crossbars. In *Proc. of ISCA*, 2016. DOI: 10.1145/3007787.3001139 74

[192] P. Chi, S. Li, Z. Qi, P. Gu, C. Xu, T. Zhang, J. Zhao, Y. Liu, Y. Wang, and Y. Xie. PRIME: A novel processing-in-memory architecture for neural network computation in ReRAM-based main memory. In *Proc. of ISCA-43*, 2016. DOI: 10.1145/3007787.3001140 74

[193] L. Song, X. Qian, H. Li, and C. Yiran. PipeLayer: A pipelined ReRAM-based accelerator for deep learning. In *Proc. of HPCA*, 2017. DOI: 10.1109/hpca.2017.55 74

[194] IBM Research Blog. IBM scientists demonstrate mixed-precision in-memory computing for the first time; hybrid design for AI hardware, 2018. https://www.ibm.com/blogs/research/2018/04/ibm-scientists-demonstrate-mixed-precision-in-memory-computing-for-the-first-time-hybrid-design-for-ai-hardware/ 74

[195] B. Feinberg, S. Wang, and E. Ipek. Making memristive neural network accelerators reliable. In *Proc. of HPCA*, 2018. DOI: 10.1109/hpca.2018.00015 74

[196] B. Feinberg, U. Vengalam, N. Whitehair, S. Wang, and E. Ipek. Enabling scientific computing on memristive accelerators. In *Proc. of ISCA*, 2018. DOI: 10.1109/isca.2018.00039 75

[197] D. Fujiki, S. Mahlke, and R. Das. In-memory data parallel processor. In *Proc. of ASPLOS*, 2018. DOI: 10.1145/3296957.3173171 75

[198] S. Kvatinsky, D. Belousov, S. Liman, G. Satat, N. Wald, E. G. Friedman, A. Kolodny, and U. C. Weiser. MAGIC—memristor-aided logic. *IEEE Transactions on Circuits and Systems II: Express Briefs*, 61(11), 2014. DOI: 10.1109/tcsii.2014.2357292 75

[199] S. Kvatinsky, N. Wald, G. Satat, A. Kolodny, U. C. Weiser, and E. G. Friedman. MRL—memristor ratioed logic. In *Proc. of Workshop on Cellular Nanoscale Networks and their Applications*, 2012. DOI: 10.1109/cnna.2012.6331426 75

[200] S. Aga, S. Jeloka, A. Subramaniyan, S. Narayanasamy, D. Blaauw, and R. Das. Compute caches. In *IEEE International Symposium on High Performance Computer Architecture (HPCA)*, pages 481–492, 2017. DOI: 10.1109/hpca.2017.21 75

[201] S. Li, C. Xu, Q. Zou, J. Zhao, Y. Lu, and Y. Xie. Pinatubo: A processing-in-memory architecture for bulk bitwise operations in emerging non-volatile memories. In *Proc. of DAC*, 2016. DOI: 10.1145/2897937.2898064 75

[202] Q. Guo, X. Guo, Y. Bai, and E. Ipek. A resistive TCAM accelerator for data-intensive computing. In *Proc. of MICRO*, 2011. DOI: 10.1145/2155620.2155660 75

[203] Q. Guo, X. Guo, R. Patel, E. Ipek, and E. G. Friedman. AC-DIMM: Associative computing with STT-MRAM. In *Proc. of ISCA*, 2013. DOI: 10.1145/2508148.2485939 75

[204] S. Jeloka, N. B. Akesh, D. Sylvester, and D. Blaauw. A 28 nm configurable memory (TCAM/BCAM/SRAM) using push-rule 6t bit cell enabling logic-in-memory. *IEEE Journal of Solid-State Circuits*, 51(4):1009–1021, 2016. DOI: 10.1109/JSSC.2016.2515510 75

[205] C. Eckert, X. Wang, J. Wang, A. Subramaniyan, R. Iyer, D. Sylvester, D. Blaauw, and R. Das. Neural cache: Bit-serial in-cache acceleration of deep neural networks. In *Proc. of ISCA*, 2018. DOI: 10.1109/isca.2018.00040 75

[206] P. Srivastava, M. Kang, S. Gonugondla, S. Lim, J. Choi, V. Adve, N. S. Kim, and N. Shanbhag. PROMISE: An end-to-end design of a programmable mixed-signal accelerator for machine-learning algorithms. In *Proc. of ISCA*, 2018. DOI: 10.1109/isca.2018.00015 76

[207] H. Kim, J. Sim, Y. Choi, and L-S. Kim. NAND-Net: Minimizing computational complexity of in-memory processing for binary neural networks. In *Proc. of HPCA*, 2019. DOI: 10.1109/hpca.2019.00017 76

[208] S. Li, D. Niu, K. Malladi, H. Zheng, B. Brennan, and Y. Xie. DRISA: A DRAM-based reconfigurable in-situ Accelerator. In *Proc. of MICRO*, 2017. DOI: 10.1145/3123939.3123977 76

[209] V. Seshadri, K. Hsieh, A. Boroumand, D. Lee, M. Kozuch, O. Mutlu, P. Gibbons, and T. Mowry. Fast bulk bitwise AND and OR in DRAM. *IEEE Computer Architecture Letters*, 2015. DOI: 10.1109/lca.2015.2434872 76

[210] S. Li, A. O. Glova, X. Hu, P. Gu, D. Niu, K. Malladi, H. Zheng, B. Brennan, and Y. Xie. SCOPE: A stochastic computing engine for DRAM-based in-situ accelerator. In *Proc. of MICRO*, 2018. DOI: 10.1109/micro.2018.00062 76, 79

[211] V. Seshadri, D. Lee, T. Mullins, H. Hassan, A. Boroumand, J. Kim, M. Kozuch, O. Mutlu, P. Gibbons, and T. Mowry. Ambit: In-memory accelerator for bulk bitwise operations using commodity DRAM technology. In *Proc. of MICRO*, 2017. DOI: 10.1145/3123939.3124544 76

[212] P. Dlugosch, D. Brown, P. Glendenning, M. Leventhal, and H. Noyes. An efficient and scalable semiconductor architecture for parallel automata processing. *IEEE Transactions on Parallel and Distributed Systems*, 25(12), 2014. DOI: 10.1109/tpds.2014.8 77

[213] K. Wang, E. Sadredini, and K. Skadron. Sequential pattern mining with the micron automata processor. In *Proc. of CF*, 2016. DOI: 10.1145/2903150.2903172 77

[214] K. Wang, Y. Qi, J. Fox, M. Stan, and K. Skadron. Association rule mining with the micron automata processor. In *Proc. of IPDPS*, 2015. DOI: 10.1109/ipdps.2015.101

[215] I. Roy. Algorithmic techniques for the micron automata processor. Ph.D. thesis, Georgia Institute of Technology, 2015. 77

[216] J. Ahn, S. Yoo, O. Mutlu, and K. Choi. PIM-enabled instructions: A low-overhead, locality-aware processing-in-memory architecture. In *Proc. of ISCA*, 2015. DOI: 10.1145/2872887.2750385 78

[217] L. Nai, R. Hadidi, J. Sim, H. Kim, P. Kumar, and H. Kim. GraphPIM: Enabling instruction-level PIM offloading in graph computing frameworks. In *Proc. of HPCA*, 2017. DOI: 10.1109/hpca.2017.54 78, 79

[218] J. Huang, R. R. Puli, P. Majumder, S. Kim, R. Boyapati, K. H. Yum, and E. J. Kim. Active-routing: Compute on the way for near-data processing. In *Proc. of HPCA*, 2019. DOI: 10.1109/hpca.2019.00018 78

[219] K. Hsieh, E. Ebrahimi, G. Kim, N. Chatterjee, M. O'Connor, N. Vijaykumar, O. Mutlu, and S. W. Keckler. Transparent offloading and mapping (TOM): Enabling programmer-transparent near-data processing in GPU systems. In *Proc. of ISCA*, 2016. DOI: 10.1145/3007787.3001159 78

[220] G. Kim, N. Chatterjee, M. O'Connor, and K. Hsieh. Toward standardized near-data processing with unrestricted data placement for GPUs. In *Proc. of SC*, 2017. DOI: 10.1145/3126908.3126965 79

[221] C. Xie, X. Zhang, A. Li, X. Fu, and S. L. Song. PIM-VR: Erasing motion anomalies in highly-interactive virtual reality world with customized memory cube. In *Proc. of HPCA*, 2019. DOI: 10.1109/hpca.2019.00013 79

[222] A. Gutierrez, M. Cieslak, B. Giridhar, R. Dreslinski, L. Ceze, and T. Mudge. Integrated 3D-stacked server designs for increasing physical density of key-value stores. In *Proc. of ASPLOS-19*, 2014. DOI: 10.1145/2541940.2541951 79

[223] S. Blanas. Near data computing from a database systems perspective, 2018. https://www.sigarch.org/near-data-computing-from-a-database-systems-perspective/ 79

[224] Y. Chen, T. Luo, S. Liu, S. Zhang, L. He, J. Wang, L. Li, T. Chen, Z. Xu, N. Sun, et al. DaDianNao: A machine-learning supercomputer. In *Proc. of MICRO-47*, 2014. DOI: 10.1109/micro.2014.58 79

[225] D. Kim, J. H. Kung, S. Chai, S. Yalamanchili, and S. Mukhopadhyay. Neurocube: A programmable digital neuromorphic architecture with high-density 3D memory. In *Proc. of ISCA-43*, 2016. DOI: 10.1109/isca.2016.41 79

[226] M. Gao, J. Pu, X. Yang, M. Horowitz, and C. Kozyrakis. TETRIS: Scalable and efficient neural network acceleration with 3D memory. In *Proc. of ASLPOS-22*, 2017. DOI: 10.1145/3093336.3037702 79

[227] J. Liu, H. Zhao, M. Ogleari, D. Li, and J. Zhao. Processing-in-memory for energy-efficient neural network training: A heterogeneous approach. In *Proc. of MICRO*, 2018. DOI: 10.1109/micro.2018.00059 79

[228] B. Hong, Y. Ro, and J. Kim. Multi-dimensional parallel training of winograd layer on memory-centric architecture. In *Proc. of MICRO*, 2018. DOI: 10.1109/micro.2018.00061 79

[229] H. Mao, M. Song, T. Li, Y. Dai, and J. Shu. LerGAN: A zero-free, low data movement and PIM-based GAN architecture. In *Proc. of MICRO*, 2018. DOI: 10.1109/micro.2018.00060 79

[230] L. Song, Y. Zhuo, X. Qian, H. Li, and Y. Chen. GraphR: Accelerating graph processing using ReRAM. In *Proc. of HPCA*, 2018. DOI: 10.1109/hpca.2018.00052 79

[231] J. Ahn, S. Hong, S. Yoo, O. Mutlu, and K. Choi. A scalable processing-in-memory accelerator for parallel graph processing. In *Proc. of ISCA*, 2015. DOI: 10.1145/2872887.2750386 79

[232] M. Zhang, Y. Zhuo, C. Wang, M. Gao, Y. Wu, K. Chen, C. Kozyrakis, and X. Qian. GraphP: Reducing communication of PIM-based graph processing with efficient data partition. In *Proc. of HPCA*, 2018. DOI: 10.1109/hpca.2018.00053 79

[233] M. Drumond, A. Daglis, N. Mirzadeh, D. Ustiugov, J. Picorel, B. Falsafi, B. Grot, and D. Pnevmatikatos. The mondrian data engine. In *Proc. of ISCA*, 2017. DOI: 10.1145/3140659.3080233 80

[234] A. Boroumand, S. Ghose, Y. Kim, R. Ausavarungnirun, E. Shiu, R. Thakur, D. Kim, A. Kuusela, A. Knies, P. Ranganathan, and O. Mutlu. Google workloads for consumer devices: Mitigating data movement bottlenecks. In *Proc. of ASPLOS*, 2018. DOI: 10.1145/3296957.3173177 80

[235] Y. Wang, A. Ferraiuolo, and G. E. Suh. Timing channel protection for a shared memory controller. In *HPCA*, 2014. DOI: 10.1109/hpca.2014.6835934 82

[236] T. Ristenpart, E. Tromer, H. Shacham, and S. Savage. Hey, you, get off of my cloud: Exploring information leakage in third-party compute clouds. In *Proc. of the 16th ACM Conference on Computer and Communications Security*, pages 199–212, 2009. DOI: 10.1145/1653662.1653687 82

[237] P. Pessl, D. Gruss, C. Maurice, M. Schwarz, and S. Mangard. DRAMA: Exploiting DRAM addressing for cross-CPU attacks. In *Proc. of USENIX Security Symposium*, 2016. 82

[238] C. Hunger, M. Kazdagli, A. Rawat, S. Vishwanath, A. Dimakis, and M. Tiwari. Understanding contention-based channels and using them for defense. In *Proc. of HPCA*, 2015. DOI: 10.1109/hpca.2015.7056069 82

[239] A. Shafiee, A. Gundu, M. Shevgoor, R. Balasubramonian, and M. Tiwari. Avoiding information leakage in the memory controller with fixed service policies. In *Proc. of MICRO*, 2015. DOI: 10.1145/2830772.2830795 83

[240] Y. Wang, B. Wu, and G. E. Suh. Secure dynamic memory scheduling against timing channel attacks. In *Proc. of HPCA*, 2017. DOI: 10.1109/hpca.2017.27 84, 87

[241] A. Vuong, A. Shafiee, M. Taassori, and R. Balasubramonian. An MLP-aware leakage-free memory controller. In *Proc. of HASP Workshop, in Conjunction with ISCA-45*, 2018. DOI: 10.1145/3214292.3214296 85

[242] A. Ferraiuolo, Y. Wang, D. Zhang, A. C. Myers, and G. E. Suh. Lattice priority scheduling: Low-overhead timing channel protection for a shared memory controller. In *Proc. of HPCA*, 2016. DOI: 10.1109/hpca.2016.7446080 87

[243] M. Islam, M. Kuzu, and M. Kantarcioglu. Access pattern disclosure on searchable encryption: Ramification, attack, and mitigation. In *Proc. of NDSS*, 2012. 87

[244] O. Goldreich. Towards a theory of software protection and simulation by oblivious RAMs. In *Proc. of STOC*, 1987. DOI: 10.1145/28395.28416 88

[245] E. Stefanov, M. van Dijk, E. Shi, C. Fletcher, L. Ren, X. Yu, and S. Devadas. Path ORAM: An extremely simple oblivious RAM protocol. In *Proc. of CCS*, 2013. DOI: 10.1145/2508859.2516660 88

[246] W. Wang, G. Chen, X. Pan, Y. Zhang, X. Wang, V. Bindschaedler, H. Tang, and C. Gunter. Leaky cauldron on the dark land: Understanding memory side-channel hazards in SGX. *ArXiv Preprint ArXiv:1705.07289*, 2017. DOI: 10.1145/3133956.3134038 88, 96

[247] J. Bulck, N. Weichbrodt, R. Kapitza, F. Piessens, and R. Strackx. Telling your secrets without page faults: Stealthy page table-based attacks on enclaved execution. In *26th USENIX Security Symposium (USENIX Security 17)*, pages 1041–1056, USENIX Association, Vancouver, BC, 2017.

[248] S. Shinde, Z. L. Chua, V. Narayanan, and P. Saxena. Preventing page faults from telling your secrets. In *Proc. of the 11th ACM on Asia Conference on Computer and Communications Security*, pages 317–328, 2016. DOI: 10.1145/2897845.2897885 88, 96

[249] S. K. Haider and M. van Dijk. Flat ORAM: A simplified write-only oblivious RAM construction for secure processors. In *ArXiv Preprint ArXiv:1611.01571*, 2017. DOI: 10.3390/cryptography3010010 88

[250] M. Maas, E. Love, E. Stefanov, M. Tiwari, E. Shi, K. Asanovic, J. Kubiatowic, and D. Song. PHANTOM: Practical oblivious computation in a secure processor. In *Proc. of CCS*, 2013. DOI: 10.1145/2508859.2516692 88, 89, 90

[251] C. Fletcher, L. Ren, A. Kwon, M. van Dijk, and S. Devadas. Freecursive ORAM: [Nearly] free recursion and integrity verification for position-based oblivious RAM. In *Proc. of ASPLOS*, 2015. DOI: 10.1145/2786763.2694353 89, 90

[252] L. Ren, X. Yu, C. Fletcher, M. van Dijk, and S. Devadas. Design space exploration and optimization of path oblivious RAM in secure processors. In *Proc. of ISCA*, 2013. DOI: 10.1145/2508148.2485971 90

[253] X. Zhang, G. Sun, C. Zhang, W. Zhang, Y. Liang, T. Wang, Y. Chen, and J. Di. Fork path: Improving efficiency of ORAM by removing redundant memory accesses. In *Proc. of the 48th International Symposium on Microarchitecture*, 2015. DOI: 10.1145/2830772.2830787 90

[254] A. Shafiee, R. Balasubramanian, M. Tiwari, and F. Li. Secure DIMM: Moving ORAM primitives closer to memory. In *Proc. of HPCA*, 2018. DOI: 10.1109/hpca.2018.00044 91

[255] C. Nagarajan, A. Shafiee, R. Balasubramanian, and M. Tiwari. ρ: Relaxed hierarchical ORAM. In *Proc. of ASPLOS*, 2019. DOI: 10.1145/3297858.3304045 91

[256] X. Zhang, G. Sun, P. Xie, C. Zhang, Y. Liu, L. Wei, Q. Xu, and C. J. Xue. Shadow block: Accelerating ORAM accesses with data duplication. In *Proc. of MICRO*, 2018. DOI: 10.1109/micro.2018.00082 91

[257] M. Seaborn and T. Dullien. Exploiting the DRAM row hammer bug to gain kernel privileges. In *Black Hat*, 2015. 92, 100

[258] B. Rogers, S. Chhabra, Y. Solihin, and M. Prvulovic. Using address independent seed encryption and bonsai merkle trees to make secure processors OS- and performance-friendly. In *Proc. of MICRO*, 2007. DOI: 10.1109/micro.2007.16 92, 93

[259] V. Costan and S. Devadas. Intel SGX explained, 2016. `https://eprint.iacr.org/2016/086.pdf` 93

[260] Intel. Intel software guard extensions programming reference. `software.intel.com/sites/default/files/329298--001.pdf`, 2013. 93

[261] S. Arnautov, B. Trach, F. Gregor, T. Knauth, A. Martin, C. Priebe, J. Lind, D. Muthuku-maran, D. O'Keeffe, M. Stillwell, et al. SCONE: Secure linux containers with intel SGX. In *OSDI*, pages 689–703, 2016. 95

[262] M. Orenbach, P. Lifshits, M. Minkin, and M. Silberstein. Eleos: ExitLess OS services for SGX enclaves. In *EuroSys*, pages 238–253, 2017. DOI: 10.1145/3064176.3064219 95

[263] M. Taassori, A. Shafiee, and R. Balasubramonian. VAULT: Reducing paging overheads in SGX with efficient integrity verification structures. In *Proc. of ASPLOS*, 2018. DOI: 10.1145/3296957.3177155 95, 96

[264] G. Saileshwar, P. J. Nair, P. Ramrakhyani, W. Elsasser, J. A. Joao, and M. K. Qureshi. Morphable counters: Enabling compact integrity trees for low-overhead secure memories. In *Proc. of MICRO*, 2018. DOI: 10.1109/micro.2018.00041 97

[265] G. Saileshwar, P. J. Nair, P. Ramrakhyani, W. Elsasser, and M. K. Qureshi. SYNERGY: Rethinking secure-memory design for error-correcting memories. In *Proc. of HPCA*, 2018. DOI: 10.1109/hpca.2018.00046 97

[266] A. Awad, Y. Wang, D. Shands, and Y. Solihin. ObfusMem: A low-overhead access obfuscation for trusted memories. In *International Symposium on Computer Architecture*, 2017. DOI: 10.1145/3079856.3080230 98

[267] S. Aga and S. Narayanasamy. InvisiMem: Smart memory for trusted computing. In *International Symposium on Computer Architecture*, 2017. 98, 99

[268] Y. Kim, R. Daly, J. Kim, C. Fallin, J. Lee, D. Lee, C. Wilkerson, K. Lai, and O. Mutlu. Flipping bits in memory without accessing them: An experimental study of DRAM disturbance errors. In *Proc. of ISCA*, 2014. DOI: 10.1109/isca.2014.6853210 99, 100

[269] V. Young, P. J. Nair, and M. K. Qureshi. DEUCE: Write-efficient encryption for non-volatile memories. In *Proc. of ASPLOS*, 2015. DOI: 10.1145/2786763.2694387 99

[270] S. Liu, A. Kolli, J. Ren, and S. Khan. Crash consistency in encrypted non-volatile main memory systems. In *Proc. of HPCA*, 2018. DOI: 10.1109/hpca.2018.00035 100

[271] M. Ye, C. Hughes, and A. Awad. Osiris: A low-cost mechanism to enable restoration of secure non-volatile memories. In *Proc. of MICRO*, 2018. DOI: 10.1109/micro.2018.00040 100

[272] P. Zuo, Y. Hua, M. Zhao, W. Zhou, and Y. Guo. Improving the performance and endurance of encrypted non-volatile main memory through deduplicating writes. In *Proc. of MICRO*, 2018. DOI: 10.1109/micro.2018.00043 100

[273] K. Bains, J. Halbert, C. Mozak, T. Schoenborn, and Z. Greenfield. Row hammer refresh command. U.S. Patent Application 2014/0059287, 2014. 100

[274] D-H. Kim, P. J. Nair, and M. K. Qureshi. Architectural support for mitigating row hammering in DRAM memories. *IEEE Computer Architecture Letters*, 2015. DOI: 10.1109/lca.2014.2332177 100

[275] S. Seyedzadeh, A. K. Jones, and R. Melhem. Mitigating wordline crosstalk using adaptive trees of counters. In *Proc. of ISCA*, 2018. DOI: 10.1109/isca.2018.00057 100

[276] Z. B. Aweke, S. F. Yitbarek, R. Qiao, R. Das, M. Hicks, Y. Oren, and T. Austin. ANVIL: Software-based protection against next-generation rowhammer attacks. In *Proc. of ASPLOS*, 2016. DOI: 10.1145/2954680.2872390 100

[277] J. L. Hennessy and D. A. Patterson. *Computer Architecture: A Quantitative Approach*, 5th ed., Elsevier, 2011. 1

Author's Biography

RAJEEV BALASUBRAMONIAN

Rajeev Balasubramonian is a Professor at the School of Computing, University of Utah. He received his B.Tech. in Computer Science and Engineering from the Indian Institute of Technology, Bombay in 1998. He received his M.S. (2000) and Ph.D. (2003) from the University of Rochester. His primary research interests include memory systems, security, and application-specific architectures. Prof. Balasubramonian is a recipient of an NSF CAREER award, faculty research awards from IBM, Google, HPE, an Intel Outstanding Research Award, and various teaching awards at the University of Utah. He has co-authored papers that have been selected as IEEE Micro Top Picks (2007 and 2010) and that have received three best paper awards.

Printed in the United States
by Baker & Taylor Publisher Services